Grades 3-4

BUILDING MATH SKILLS

BY Brian Rhee

Over 2000 Practice Problems

Detailed Solutions

Legal Notice

Published by: Solomon Academy
First Edition
ISBN-13: 978-1717064189
ISBN-10: 1717064183

About This Book

This book is designed to help students build basic arithmetic and math skills. There are resources pages in the beginning of the book that illustrate how to do each type of problems. It is strongly recommended to go over the resources pages before solving practice problems.

This book contains 12 lessons with detailed solutions. Each lesson has ten practice worksheets which provides challenges to improve and strengthen students' math skills. Completing 12 lessons enables students to build their confidence and master their math skills.

About Author

Brian(Yeon) Rhee obtained a Masters of Arts Degree in Statistics at Columbia University, NY. He served as the Mathematical Statistician at the Bureau of Labor Statistics, DC. He is the Head Academic Director at Solomon Academy due to his devotion to the community coupled with his passion for teaching. His mission is to help students of all confidence level excel in academia to build a strong foundation in character, knowledge, and wisdom. Now, Solomon academy is known as the best academy specialized in Math in Northern Virginia.

Brian Rhee has published more than ten books. The titles of his books are 7 full-length practice tests for the AP Calculus AB/BC Multiple choice sections, AP Calculus, SAT 1 Math, SAT 2 Math level 2, 12 full-length practice tests for the SAT 2 Math Level 2, SHSAT/TJHSST Math workbook, and IAAT (Iowa Algebra Aptitude Test) Volume 1 and 2, CogAT form 7 Level 8, and NNAT 2 Level B Grade 1. He's currently working on other math books which will be introduced in the near future.

Brian Rhee has more than twenty years of teaching experience in math. He has been one of the most popular tutors among TJHSST (Thomas Jefferson High School For Science and Technology) students. Currently, he is developing many online math courses with www.masterprep.net for AP Calculus AB and BC, SAT 2 Math level 2 test, and other various math subjects.

SOLOMON ACADEMY

Solomon Academy is a prestigious institution of learning with numerous qualified teachers of various fields of education. Our mission is to thoroughly teach students of all ages and confidence levels, elevate skills to the highest standard of education, and provide them with all the tools and materials to succeed.

5723 Centre Square Drive
Centreville, VA 20120
Tel: 703-988-0019

Email: solomonacademyva@gmail.com
info@solomonacademy.net

CLASSES OFFERED

MATHEMATICS	TESTING	ENGLISH
1st-6th grade math	CogAt	1st-6th Reading
Algebra 1, 2	IAAT and SOL 7	1st-6th Writing
Geometry	TJHSST Prep	Essay Writing
Pre-Calculus	SAT/ACT Prep	SAT Writing
AP Calculus AB BC	SAT 2 Subject Tests	
AP Statistics	MathCounts	
Multivariate Calculus	AMC 10/12	

LEARN FROM THE AUTHOR

Private sessions with Brian Rhee is also available on the following subjects: SAT Math, SAT 2 Subject Math Level 2, Pre-Calculus, AP Calculus AB/BC, AP Statistics, IB SL/HL, Multivariate Calculus, Linear Algebra, AMC 8/10/12, and AIME.

Feel free to contact me at solomonacademyva@gmail.com

Acknowledgements

I wish to acknowledge my deepest appreciation to my wife, Sookyung, who has continuously given me wholehearted support, encouragement, and love. Without you, I could not have completed this book.

Thank you to my sons, Joshua and Jason, who have given me big smiles and inspiration. I love you all.

Thank you to Mr. Kwon from www.Masterprep.net, who has given me opportunities to develop online math courses for various math subjects.

Contents

How to do multiplication and division?

- Multiplication without regrouping

  1. Multiply 1×3.
  2. Multiply 2×3.
  3. Multiply 3×3.

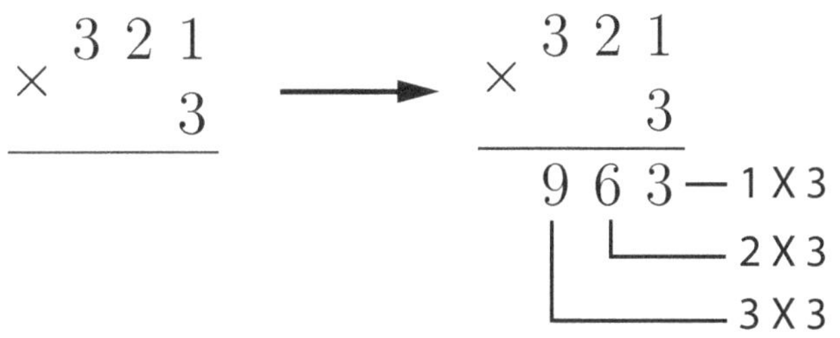

$$
\begin{array}{r}
3\ 2\ 1 \\
\times \quad 3 \\
\end{array}
\longrightarrow
\begin{array}{r}
3\ 2\ 1 \\
\times \quad 3 \\
\hline
9\ 6\ 3
\end{array}
$$

9 6 3 — 1 X 3
— 2 X 3
— 3 X 3

- Multiplication with regrouping

  1. Multiply 47×6.
  2. Multiply 47×5.
  3. Add.

$$
\begin{array}{r}
4\ 7 \\
\times\ 5\ 6 \\
\end{array}
\longrightarrow
$$

$$
\begin{array}{r}
4\ 7 \\
\times\ 5\ 6 \\
\hline
2\ 8\ 2 \\
+\ 2\ 3\ 5 \\
\hline
2\ 6\ 3\ 2
\end{array}
$$

47 X 6
47 X 5

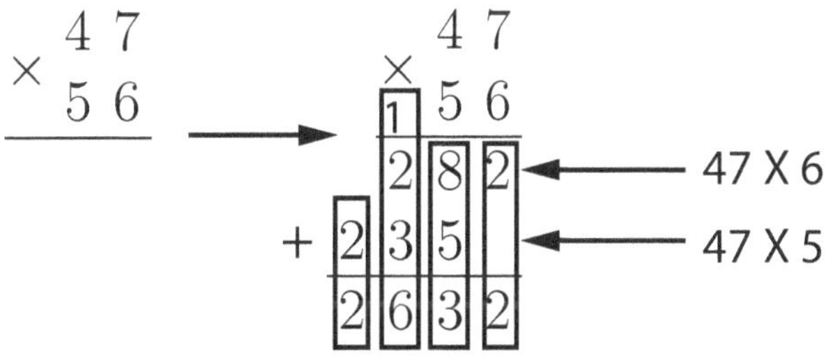

- Division without a remainder

  1. 3 goes into 4 one time.
  2. Subtract 3 from 4.
  3. Bring down the 2. 3 goes into 12 four times.

$$3\overline{)42} \longrightarrow \begin{array}{r} 14 \\ 3\overline{)42} \\ \underline{30} \\ 12 \\ \underline{12} \\ 0 \end{array}$$

- Division with a remainder

  1. 5 goes into 6 one time.
  2. Subtract 5 from 6.
  3. Bring down the 7. 5 goes into 17 three times.
  4. The remainder is 2.

$$5\overline{)67} \longrightarrow \begin{array}{r} 13 \\ 5\overline{)67} \\ \underline{50} \\ 17 \\ \underline{15} \\ 2 \end{array}$$

Order of operations

To simplify numerical expressions, use the **order of operations** (PEMDAS)

- P: Parenthesis
- E: Exponent(Ignore)
- M: Multiplication
- D: Division
- A: Addition
- S: Subtraction

The order of operations suggests to first perform any calculations inside parentheses. Afterwards, evaluate any exponents(Ignore). Next, perform all multiplications and divisions working from left to right. Finally, do additions and subtractions from left to right. For example,

$$\begin{aligned}(31-16)-15\div 5 &= 15-15\div 5 && \text{Perform subtraction inside parentheses.}\\ &= 15-3 && \text{Perform division}\\ &= 12\end{aligned}$$

Name: **Lesson 1-1 Multiplying two digit numbers**

1. $\begin{array}{r} 87 \\ \times\ 22 \\ \hline \end{array}$

2. $\begin{array}{r} 34 \\ \times\ 63 \\ \hline \end{array}$

3. $\begin{array}{r} 39 \\ \times\ 42 \\ \hline \end{array}$

4. $\begin{array}{r} 69 \\ \times\ 52 \\ \hline \end{array}$

5. $\begin{array}{r} 39 \\ \times\ 63 \\ \hline \end{array}$

6. $\begin{array}{r} 67 \\ \times\ 32 \\ \hline \end{array}$

7. $\begin{array}{r} 97 \\ \times\ 79 \\ \hline \end{array}$

8. $\begin{array}{r} 48 \\ \times\ 83 \\ \hline \end{array}$

9. $\begin{array}{r} 77 \\ \times\ 28 \\ \hline \end{array}$

10. $\begin{array}{r} 48 \\ \times\ 72 \\ \hline \end{array}$

11. $\begin{array}{r} 35 \\ \times\ 44 \\ \hline \end{array}$

12. $\begin{array}{r} 98 \\ \times\ 55 \\ \hline \end{array}$

13. $\begin{array}{r} 85 \\ \times\ 64 \\ \hline \end{array}$

14. $\begin{array}{r} 47 \\ \times\ 34 \\ \hline \end{array}$

15. $\begin{array}{r} 74 \\ \times\ 84 \\ \hline \end{array}$

16. $\begin{array}{r} 99 \\ \times\ 98 \\ \hline \end{array}$

Name: **Lesson 1-2 Multiplying two digit numbers**

---

1. $\begin{array}{r} 98 \\ \times\ 22 \\ \hline \end{array}$

2. $\begin{array}{r} 79 \\ \times\ 32 \\ \hline \end{array}$

3. $\begin{array}{r} 61 \\ \times\ 42 \\ \hline \end{array}$

4. $\begin{array}{r} 89 \\ \times\ 52 \\ \hline \end{array}$

5. $\begin{array}{r} 79 \\ \times\ 62 \\ \hline \end{array}$

6. $\begin{array}{r} 68 \\ \times\ 72 \\ \hline \end{array}$

7. $\begin{array}{r} 99 \\ \times\ 86 \\ \hline \end{array}$

8. $\begin{array}{r} 57 \\ \times\ 82 \\ \hline \end{array}$

9. $\begin{array}{r} 88 \\ \times\ 28 \\ \hline \end{array}$

10. $\begin{array}{r} 62 \\ \times\ 34 \\ \hline \end{array}$

11. $\begin{array}{r} 85 \\ \times\ 47 \\ \hline \end{array}$

12. $\begin{array}{r} 48 \\ \times\ 56 \\ \hline \end{array}$

13. $\begin{array}{r} 47 \\ \times\ 65 \\ \hline \end{array}$

14. $\begin{array}{r} 53 \\ \times\ 75 \\ \hline \end{array}$

15. $\begin{array}{r} 49 \\ \times\ 77 \\ \hline \end{array}$

16. $\begin{array}{r} 96 \\ \times\ 96 \\ \hline \end{array}$

Name: **Lesson 1-3 Multiplying two digit numbers**

1. $\begin{array}{r} 92 \\ \times\ 23 \\ \hline \end{array}$

2. $\begin{array}{r} 98 \\ \times\ 32 \\ \hline \end{array}$

3. $\begin{array}{r} 46 \\ \times\ 53 \\ \hline \end{array}$

4. $\begin{array}{r} 72 \\ \times\ 37 \\ \hline \end{array}$

5. $\begin{array}{r} 94 \\ \times\ 62 \\ \hline \end{array}$

6. $\begin{array}{r} 85 \\ \times\ 81 \\ \hline \end{array}$

7. $\begin{array}{r} 49 \\ \times\ 95 \\ \hline \end{array}$

8. $\begin{array}{r} 44 \\ \times\ 33 \\ \hline \end{array}$

9. $\begin{array}{r} 67 \\ \times\ 29 \\ \hline \end{array}$

10. $\begin{array}{r} 79 \\ \times\ 43 \\ \hline \end{array}$

11. $\begin{array}{r} 58 \\ \times\ 67 \\ \hline \end{array}$

12. $\begin{array}{r} 39 \\ \times\ 97 \\ \hline \end{array}$

13. $\begin{array}{r} 58 \\ \times\ 66 \\ \hline \end{array}$

14. $\begin{array}{r} 85 \\ \times\ 75 \\ \hline \end{array}$

15. $\begin{array}{r} 94 \\ \times\ 99 \\ \hline \end{array}$

16. $\begin{array}{r} 73 \\ \times\ 76 \\ \hline \end{array}$

Name: **Lesson 1-4 Multiplying two digit numbers**

1. $\begin{array}{r} 93 \\ \times\ 23 \\ \hline \end{array}$

2. $\begin{array}{r} 35 \\ \times\ 45 \\ \hline \end{array}$

3. $\begin{array}{r} 42 \\ \times\ 76 \\ \hline \end{array}$

4. $\begin{array}{r} 89 \\ \times\ 53 \\ \hline \end{array}$

5. $\begin{array}{r} 45 \\ \times\ 73 \\ \hline \end{array}$

6. $\begin{array}{r} 96 \\ \times\ 72 \\ \hline \end{array}$

7. $\begin{array}{r} 98 \\ \times\ 89 \\ \hline \end{array}$

8. $\begin{array}{r} 43 \\ \times\ 92 \\ \hline \end{array}$

9. $\begin{array}{r} 77 \\ \times\ 39 \\ \hline \end{array}$

10. $\begin{array}{r} 81 \\ \times\ 34 \\ \hline \end{array}$

11. $\begin{array}{r} 64 \\ \times\ 65 \\ \hline \end{array}$

12. $\begin{array}{r} 74 \\ \times\ 59 \\ \hline \end{array}$

13. $\begin{array}{r} 64 \\ \times\ 74 \\ \hline \end{array}$

14. $\begin{array}{r} 94 \\ \times\ 73 \\ \hline \end{array}$

15. $\begin{array}{r} 76 \\ \times\ 88 \\ \hline \end{array}$

16. $\begin{array}{r} 95 \\ \times\ 87 \\ \hline \end{array}$

Name: **Lesson 1-5 Multiplying two digit numbers**

1. $\begin{array}{r} 97 \\ \times\ 33 \\ \hline \end{array}$

2. $\begin{array}{r} 37 \\ \times\ 42 \\ \hline \end{array}$

3. $\begin{array}{r} 99 \\ \times\ 42 \\ \hline \end{array}$

4. $\begin{array}{r} 91 \\ \times\ 55 \\ \hline \end{array}$

5. $\begin{array}{r} 59 \\ \times\ 63 \\ \hline \end{array}$

6. $\begin{array}{r} 82 \\ \times\ 72 \\ \hline \end{array}$

7. $\begin{array}{r} 98 \\ \times\ 83 \\ \hline \end{array}$

8. $\begin{array}{r} 58 \\ \times\ 99 \\ \hline \end{array}$

9. $\begin{array}{r} 89 \\ \times\ 37 \\ \hline \end{array}$

10. $\begin{array}{r} 95 \\ \times\ 43 \\ \hline \end{array}$

11. $\begin{array}{r} 36 \\ \times\ 63 \\ \hline \end{array}$

12. $\begin{array}{r} 84 \\ \times\ 65 \\ \hline \end{array}$

13. $\begin{array}{r} 78 \\ \times\ 65 \\ \hline \end{array}$

14. $\begin{array}{r} 86 \\ \times\ 97 \\ \hline \end{array}$

15. $\begin{array}{r} 74 \\ \times\ 84 \\ \hline \end{array}$

16. $\begin{array}{r} 89 \\ \times\ 89 \\ \hline \end{array}$

Name: **Lesson 1-6 Multiplying two digit numbers**

1. $\begin{array}{r} 96 \\ \times\ 39 \\ \hline \end{array}$

2. $\begin{array}{r} 74 \\ \times\ 33 \\ \hline \end{array}$

3. $\begin{array}{r} 52 \\ \times\ 43 \\ \hline \end{array}$

4. $\begin{array}{r} 49 \\ \times\ 54 \\ \hline \end{array}$

5. $\begin{array}{r} 78 \\ \times\ 63 \\ \hline \end{array}$

6. $\begin{array}{r} 52 \\ \times\ 73 \\ \hline \end{array}$

7. $\begin{array}{r} 97 \\ \times\ 83 \\ \hline \end{array}$

8. $\begin{array}{r} 64 \\ \times\ 98 \\ \hline \end{array}$

9. $\begin{array}{r} 79 \\ \times\ 38 \\ \hline \end{array}$

10. $\begin{array}{r} 59 \\ \times\ 36 \\ \hline \end{array}$

11. $\begin{array}{r} 57 \\ \times\ 46 \\ \hline \end{array}$

12. $\begin{array}{r} 57 \\ \times\ 57 \\ \hline \end{array}$

13. $\begin{array}{r} 93 \\ \times\ 65 \\ \hline \end{array}$

14. $\begin{array}{r} 95 \\ \times\ 78 \\ \hline \end{array}$

15. $\begin{array}{r} 88 \\ \times\ 76 \\ \hline \end{array}$

16. $\begin{array}{r} 89 \\ \times\ 96 \\ \hline \end{array}$

Name: **Lesson 1-7 Multiplying two digit numbers**

1. $\begin{array}{r} 98 \\ \times\ 34 \\ \hline \end{array}$

2. $\begin{array}{r} 78 \\ \times\ 36 \\ \hline \end{array}$

3. $\begin{array}{r} 72 \\ \times\ 43 \\ \hline \end{array}$

4. $\begin{array}{r} 75 \\ \times\ 53 \\ \hline \end{array}$

5. $\begin{array}{r} 94 \\ \times\ 63 \\ \hline \end{array}$

6. $\begin{array}{r} 59 \\ \times\ 73 \\ \hline \end{array}$

7. $\begin{array}{r} 87 \\ \times\ 81 \\ \hline \end{array}$

8. $\begin{array}{r} 68 \\ \times\ 99 \\ \hline \end{array}$

9. $\begin{array}{r} 99 \\ \times\ 32 \\ \hline \end{array}$

10. $\begin{array}{r} 39 \\ \times\ 36 \\ \hline \end{array}$

11. $\begin{array}{r} 77 \\ \times\ 44 \\ \hline \end{array}$

12. $\begin{array}{r} 56 \\ \times\ 59 \\ \hline \end{array}$

13. $\begin{array}{r} 95 \\ \times\ 66 \\ \hline \end{array}$

14. $\begin{array}{r} 87 \\ \times\ 89 \\ \hline \end{array}$

15. $\begin{array}{r} 86 \\ \times\ 85 \\ \hline \end{array}$

16. $\begin{array}{r} 76 \\ \times\ 91 \\ \hline \end{array}$

Name: **Lesson 1-8 Multiplying two digit numbers**

1. $\begin{array}{r} 85 \\ \times\ 42 \\ \hline \end{array}$

2. $\begin{array}{r} 49 \\ \times\ 57 \\ \hline \end{array}$

3. $\begin{array}{r} 73 \\ \times\ 25 \\ \hline \end{array}$

4. $\begin{array}{r} 79 \\ \times\ 36 \\ \hline \end{array}$

5. $\begin{array}{r} 97 \\ \times\ 27 \\ \hline \end{array}$

6. $\begin{array}{r} 85 \\ \times\ 72 \\ \hline \end{array}$

7. $\begin{array}{r} 64 \\ \times\ 85 \\ \hline \end{array}$

8. $\begin{array}{r} 54 \\ \times\ 49 \\ \hline \end{array}$

9. $\begin{array}{r} 83 \\ \times\ 56 \\ \hline \end{array}$

10. $\begin{array}{r} 48 \\ \times\ 63 \\ \hline \end{array}$

11. $\begin{array}{r} 78 \\ \times\ 65 \\ \hline \end{array}$

12. $\begin{array}{r} 96 \\ \times\ 85 \\ \hline \end{array}$

13. $\begin{array}{r} 97 \\ \times\ 77 \\ \hline \end{array}$

14. $\begin{array}{r} 56 \\ \times\ 77 \\ \hline \end{array}$

15. $\begin{array}{r} 89 \\ \times\ 98 \\ \hline \end{array}$

16. $\begin{array}{r} 86 \\ \times\ 39 \\ \hline \end{array}$

Name: **Lesson 1-9 Multiplying two digit numbers**

1. $\begin{array}{r} 99 \\ \times\ 33 \\ \hline \end{array}$

2. $\begin{array}{r} 69 \\ \times\ 24 \\ \hline \end{array}$

3. $\begin{array}{r} 87 \\ \times\ 43 \\ \hline \end{array}$

4. $\begin{array}{r} 94 \\ \times\ 54 \\ \hline \end{array}$

5. $\begin{array}{r} 43 \\ \times\ 64 \\ \hline \end{array}$

6. $\begin{array}{r} 92 \\ \times\ 38 \\ \hline \end{array}$

7. $\begin{array}{r} 72 \\ \times\ 39 \\ \hline \end{array}$

8. $\begin{array}{r} 65 \\ \times\ 95 \\ \hline \end{array}$

9. $\begin{array}{r} 89 \\ \times\ 37 \\ \hline \end{array}$

10. $\begin{array}{r} 84 \\ \times\ 47 \\ \hline \end{array}$

11. $\begin{array}{r} 85 \\ \times\ 84 \\ \hline \end{array}$

12. $\begin{array}{r} 85 \\ \times\ 69 \\ \hline \end{array}$

13. $\begin{array}{r} 87 \\ \times\ 85 \\ \hline \end{array}$

14. $\begin{array}{r} 97 \\ \times\ 75 \\ \hline \end{array}$

15. $\begin{array}{r} 91 \\ \times\ 99 \\ \hline \end{array}$

16. $\begin{array}{r} 87 \\ \times\ 98 \\ \hline \end{array}$

Name: **Lesson 1-10 Multiplying two digit numbers**

1. $\begin{array}{r} 76 \\ \times\ 27 \\ \hline \end{array}$

2. $\begin{array}{r} 53 \\ \times\ 59 \\ \hline \end{array}$

3. $\begin{array}{r} 76 \\ \times\ 86 \\ \hline \end{array}$

4. $\begin{array}{r} 57 \\ \times\ 51 \\ \hline \end{array}$

5. $\begin{array}{r} 65 \\ \times\ 63 \\ \hline \end{array}$

6. $\begin{array}{r} 85 \\ \times\ 55 \\ \hline \end{array}$

7. $\begin{array}{r} 71 \\ \times\ 71 \\ \hline \end{array}$

8. $\begin{array}{r} 74 \\ \times\ 93 \\ \hline \end{array}$

9. $\begin{array}{r} 79 \\ \times\ 81 \\ \hline \end{array}$

10. $\begin{array}{r} 87 \\ \times\ 74 \\ \hline \end{array}$

11. $\begin{array}{r} 86 \\ \times\ 94 \\ \hline \end{array}$

12. $\begin{array}{r} 56 \\ \times\ 88 \\ \hline \end{array}$

13. $\begin{array}{r} 75 \\ \times\ 79 \\ \hline \end{array}$

14. $\begin{array}{r} 62 \\ \times\ 97 \\ \hline \end{array}$

15. $\begin{array}{r} 94 \\ \times\ 78 \\ \hline \end{array}$

16. $\begin{array}{r} 97 \\ \times\ 66 \\ \hline \end{array}$

Name:

Lesson 2-1 Dividing by one digit numbers with remainders

1. $2\,\overline{)43}$	2. $2\,\overline{)63}$	3. $3\,\overline{)59}$	4. $3\,\overline{)76}$
5. $4\,\overline{)39}$	6. $5\,\overline{)72}$	7. $5\,\overline{)88}$	8. $4\,\overline{)95}$
9. $3\,\overline{)50}$	10. $2\,\overline{)35}$	11. $3\,\overline{)49}$	12. $3\,\overline{)92}$
13. $3\,\overline{)70}$	14. $5\,\overline{)99}$	15. $6\,\overline{)83}$	16. $5\,\overline{)69}$
17. $4\,\overline{)51}$	18. $4\,\overline{)98}$	19. $6\,\overline{)63}$	20. $4\,\overline{)61}$

Name: **Lesson 2-2 Dividing by one digit numbers with remainders**

1. $2\,\overline{)37}$
2. $5\,\overline{)74}$
3. $3\,\overline{)52}$
4. $4\,\overline{)61}$
5. $4\,\overline{)97}$
6. $2\,\overline{)75}$
7. $5\,\overline{)99}$
8. $8\,\overline{)99}$
9. $2\,\overline{)43}$
10. $4\,\overline{)78}$
11. $3\,\overline{)70}$
12. $3\,\overline{)79}$
13. $3\,\overline{)37}$
14. $2\,\overline{)87}$
15. $6\,\overline{)77}$
16. $4\,\overline{)69}$
17. $5\,\overline{)66}$
18. $5\,\overline{)82}$
19. $3\,\overline{)85}$
20. $2\,\overline{)75}$

Name: **Lesson 2-3 Dividing by one digit numbers with remainders**

1. $2\overline{)31}$ 2. $5\overline{)78}$ 3. $3\overline{)61}$ 4. $6\overline{)99}$

5. $4\overline{)91}$ 6. $3\overline{)89}$ 7. $6\overline{)87}$ 8. $3\overline{)95}$

9. $2\overline{)39}$ 10. $4\overline{)83}$ 11. $3\overline{)89}$ 12. $7\overline{)76}$

13. $5\overline{)57}$ 14. $3\overline{)53}$ 15. $6\overline{)88}$ 16. $3\overline{)64}$

17. $2\overline{)79}$ 18. $6\overline{)64}$ 19. $3\overline{)76}$ 20. $2\overline{)41}$

Name: **Lesson 2-4 Dividing by one digit numbers with remainders**

1. $2\overline{)59}$	2. $2\overline{)85}$	3. $6\overline{)67}$	4. $7\overline{)99}$
5. $4\overline{)87}$	6. $5\overline{)87}$	7. $3\overline{)61}$	8. $4\overline{)97}$
9. $2\overline{)73}$	10. $2\overline{)99}$	11. $5\overline{)78}$	12. $7\overline{)78}$
13. $5\overline{)71}$	14. $6\overline{)95}$	15. $3\overline{)55}$	16. $3\overline{)98}$
17. $3\overline{)58}$	18. $2\overline{)81}$	19. $5\overline{)69}$	20. $4\overline{)65}$

Name:

Lesson 2-5 Dividing by one digit numbers with remainders

1. $3\overline{)47}$	2. $6\overline{)86}$	3. $3\overline{)77}$	4. $7\overline{)96}$
5. $5\overline{)62}$	6. $3\overline{)38}$	7. $6\overline{)88}$	8. $4\overline{)77}$
9. $3\overline{)58}$	10. $5\overline{)93}$	11. $4\overline{)46}$	12. $2\overline{)57}$
13. $5\overline{)61}$	14. $4\overline{)65}$	15. $7\overline{)87}$	16. $7\overline{)81}$
17. $2\overline{)87}$	18. $6\overline{)89}$	19. $4\overline{)55}$	20. $4\overline{)67}$

Name: **Lesson 2-6 Dividing by one digit numbers with remainders**

1. $2\overline{)33}$ 2. $5\overline{)88}$ 3. $3\overline{)71}$ 4. $7\overline{)79}$

5. $5\overline{)64}$ 6. $3\overline{)43}$ 7. $6\overline{)92}$ 8. $4\overline{)79}$

9. $3\overline{)64}$ 10. $4\overline{)83}$ 11. $6\overline{)85}$ 12. $2\overline{)57}$

13. $4\overline{)78}$ 14. $3\overline{)77}$ 15. $6\overline{)92}$ 16. $3\overline{)58}$

17. $2\overline{)97}$ 18. $6\overline{)95}$ 19. $3\overline{)55}$ 20. $6\overline{)71}$

Name: **Lesson 2-7 Dividing by one digit numbers with remainders**

---

1. $2\overline{)43}$	2. $3\overline{)35}$	3. $6\overline{)85}$	4. $4\overline{)47}$
5. $5\overline{)51}$	6. $6\overline{)76}$	7. $3\overline{)79}$	8. $7\overline{)92}$
9. $3\overline{)52}$	10. $4\overline{)54}$	11. $6\overline{)79}$	12. $5\overline{)58}$
13. $3\overline{)65}$	14. $5\overline{)97}$	15. $8\overline{)91}$	16. $8\overline{)99}$
17. $5\overline{)99}$	18. $3\overline{)79}$	19. $3\overline{)92}$	20. $7\overline{)87}$

Name:

Lesson 2-8 Dividing by one digit numbers with remainders

---

1. $2\overline{)51}$
2. $3\overline{)47}$
3. $6\overline{)89}$
4. $8\overline{)90}$
5. $5\overline{)64}$
6. $5\overline{)89}$
7. $3\overline{)83}$
8. $4\overline{)93}$
9. $2\overline{)63}$
10. $3\overline{)88}$
11. $4\overline{)50}$
12. $3\overline{)38}$
13. $6\overline{)73}$
14. $5\overline{)96}$
15. $5\overline{)84}$
16. $4\overline{)89}$
17. $4\overline{)51}$
18. $3\overline{)73}$
19. $3\overline{)67}$
20. $3\overline{)59}$

Name: **Lesson 2-9 Dividing by one digit numbers with remainders**

1. $3\overline{)55}$ 2. $3\overline{)91}$ 3. $4\overline{)75}$ 4. $6\overline{)72}$

5. $6\overline{)68}$ 6. $7\overline{)90}$ 7. $4\overline{)98}$ 8. $5\overline{)78}$

9. $3\overline{)67}$ 10. $7\overline{)79}$ 11. $8\overline{)97}$ 12. $5\overline{)67}$

13. $6\overline{)69}$ 14. $3\overline{)95}$ 15. $4\overline{)61}$ 16. $5\overline{)89}$

17. $3\overline{)76}$ 18. $4\overline{)75}$ 19. $3\overline{)83}$ 20. $3\overline{)52}$

Name: **Lesson 2-10 Dividing by one digit numbers with remainders**

1. $4\overline{)57}$	2. $4\overline{)90}$	3. $5\overline{)99}$	4. $4\overline{)63}$
5. $6\overline{)86}$	6. $4\overline{)94}$	7. $3\overline{)85}$	8. $9\overline{)98}$
9. $7\overline{)89}$	10. $4\overline{)49}$	11. $7\overline{)95}$	12. $5\overline{)99}$
13. $2\overline{)75}$	14. $5\overline{)97}$	15. $5\overline{)93}$	16. $4\overline{)67}$
17. $5\overline{)81}$	18. $5\overline{)87}$	19. $7\overline{)82}$	20. $6\overline{)77}$

Name: **Lesson 3-1 Multiplying two and three digit numbers**

---

1. $\begin{array}{r} 124 \\ \times \quad 58 \\ \hline \end{array}$

2. $\begin{array}{r} 285 \\ \times \quad 26 \\ \hline \end{array}$

3. $\begin{array}{r} 336 \\ \times \quad 25 \\ \hline \end{array}$

4. $\begin{array}{r} 427 \\ \times \quad 49 \\ \hline \end{array}$

5. $\begin{array}{r} 412 \\ \times \quad 74 \\ \hline \end{array}$

6. $\begin{array}{r} 228 \\ \times \quad 36 \\ \hline \end{array}$

7. $\begin{array}{r} 315 \\ \times \quad 56 \\ \hline \end{array}$

8. $\begin{array}{r} 524 \\ \times \quad 53 \\ \hline \end{array}$

9. $\begin{array}{r} 468 \\ \times \quad 43 \\ \hline \end{array}$

10. $\begin{array}{r} 413 \\ \times \quad 37 \\ \hline \end{array}$

11. $\begin{array}{r} 257 \\ \times \quad 46 \\ \hline \end{array}$

12. $\begin{array}{r} 396 \\ \times \quad 29 \\ \hline \end{array}$

Name: **Lesson 3-2 Multiplying two and three digit numbers**

1. $\begin{array}{r} 337 \\ \times \quad 45 \\ \hline \end{array}$

2. $\begin{array}{r} 513 \\ \times \quad 42 \\ \hline \end{array}$

3. $\begin{array}{r} 535 \\ \times \quad 48 \\ \hline \end{array}$

4. $\begin{array}{r} 442 \\ \times \quad 53 \\ \hline \end{array}$

5. $\begin{array}{r} 542 \\ \times \quad 35 \\ \hline \end{array}$

6. $\begin{array}{r} 299 \\ \times \quad 57 \\ \hline \end{array}$

7. $\begin{array}{r} 583 \\ \times \quad 42 \\ \hline \end{array}$

8. $\begin{array}{r} 458 \\ \times \quad 55 \\ \hline \end{array}$

9. $\begin{array}{r} 319 \\ \times \quad 43 \\ \hline \end{array}$

10. $\begin{array}{r} 359 \\ \times \quad 38 \\ \hline \end{array}$

11. $\begin{array}{r} 478 \\ \times \quad 57 \\ \hline \end{array}$

12. $\begin{array}{r} 456 \\ \times \quad 54 \\ \hline \end{array}$

Name:

Lesson 3-3 Multiplying two and three digit numbers

1. $$\begin{array}{r} 443 \\ \times \quad 35 \\ \hline \end{array}$$

2. $$\begin{array}{r} 524 \\ \times \quad 31 \\ \hline \end{array}$$

3. $$\begin{array}{r} 466 \\ \times \quad 36 \\ \hline \end{array}$$

4. $$\begin{array}{r} 578 \\ \times \quad 46 \\ \hline \end{array}$$

5. $$\begin{array}{r} 569 \\ \times \quad 47 \\ \hline \end{array}$$

6. $$\begin{array}{r} 396 \\ \times \quad 42 \\ \hline \end{array}$$

7. $$\begin{array}{r} 496 \\ \times \quad 37 \\ \hline \end{array}$$

8. $$\begin{array}{r} 436 \\ \times \quad 33 \\ \hline \end{array}$$

9. $$\begin{array}{r} 569 \\ \times \quad 49 \\ \hline \end{array}$$

10. $$\begin{array}{r} 354 \\ \times \quad 42 \\ \hline \end{array}$$

11. $$\begin{array}{r} 451 \\ \times \quad 48 \\ \hline \end{array}$$

12. $$\begin{array}{r} 341 \\ \times \quad 57 \\ \hline \end{array}$$

Name: **Lesson 3-4 Multiplying two and three digit numbers**

1. $\begin{array}{r} 426 \\ \times \quad 53 \\ \hline \end{array}$

2. $\begin{array}{r} 465 \\ \times \quad 53 \\ \hline \end{array}$

3. $\begin{array}{r} 563 \\ \times \quad 51 \\ \hline \end{array}$

4. $\begin{array}{r} 563 \\ \times \quad 45 \\ \hline \end{array}$

5. $\begin{array}{r} 597 \\ \times \quad 48 \\ \hline \end{array}$

6. $\begin{array}{r} 684 \\ \times \quad 48 \\ \hline \end{array}$

7. $\begin{array}{r} 674 \\ \times \quad 48 \\ \hline \end{array}$

8. $\begin{array}{r} 574 \\ \times \quad 59 \\ \hline \end{array}$

9. $\begin{array}{r} 676 \\ \times \quad 56 \\ \hline \end{array}$

10. $\begin{array}{r} 536 \\ \times \quad 56 \\ \hline \end{array}$

11. $\begin{array}{r} 657 \\ \times \quad 44 \\ \hline \end{array}$

12. $\begin{array}{r} 593 \\ \times \quad 47 \\ \hline \end{array}$

Name: **Lesson 3-5 Multiplying two and three digit numbers**

1. $\begin{array}{r} 335 \\ \times \quad 57 \\ \hline \end{array}$

2. $\begin{array}{r} 546 \\ \times \quad 51 \\ \hline \end{array}$

3. $\begin{array}{r} 547 \\ \times \quad 36 \\ \hline \end{array}$

4. $\begin{array}{r} 562 \\ \times \quad 48 \\ \hline \end{array}$

5. $\begin{array}{r} 429 \\ \times \quad 44 \\ \hline \end{array}$

6. $\begin{array}{r} 431 \\ \times \quad 57 \\ \hline \end{array}$

7. $\begin{array}{r} 578 \\ \times \quad 39 \\ \hline \end{array}$

8. $\begin{array}{r} 555 \\ \times \quad 55 \\ \hline \end{array}$

9. $\begin{array}{r} 347 \\ \times \quad 46 \\ \hline \end{array}$

10. $\begin{array}{r} 473 \\ \times \quad 45 \\ \hline \end{array}$

11. $\begin{array}{r} 469 \\ \times \quad 47 \\ \hline \end{array}$

12. $\begin{array}{r} 562 \\ \times \quad 54 \\ \hline \end{array}$

Name: **Lesson 3-6 Multiplying two and three digit numbers**

1. $\begin{array}{r} 386 \\ \times \quad 52 \\ \hline \end{array}$

2. $\begin{array}{r} 556 \\ \times \quad 65 \\ \hline \end{array}$

3. $\begin{array}{r} 587 \\ \times \quad 51 \\ \hline \end{array}$

4. $\begin{array}{r} 637 \\ \times \quad 45 \\ \hline \end{array}$

5. $\begin{array}{r} 438 \\ \times \quad 57 \\ \hline \end{array}$

6. $\begin{array}{r} 473 \\ \times \quad 42 \\ \hline \end{array}$

7. $\begin{array}{r} 528 \\ \times \quad 63 \\ \hline \end{array}$

8. $\begin{array}{r} 687 \\ \times \quad 68 \\ \hline \end{array}$

9. $\begin{array}{r} 682 \\ \times \quad 54 \\ \hline \end{array}$

10. $\begin{array}{r} 696 \\ \times \quad 64 \\ \hline \end{array}$

11. $\begin{array}{r} 398 \\ \times \quad 59 \\ \hline \end{array}$

12. $\begin{array}{r} 541 \\ \times \quad 69 \\ \hline \end{array}$

Name: **Lesson 3-7 Multiplying two and three digit numbers**

1. $\begin{array}{r} 563 \\ \times \quad 64 \\ \hline \end{array}$

2. $\begin{array}{r} 489 \\ \times \quad 28 \\ \hline \end{array}$

3. $\begin{array}{r} 667 \\ \times \quad 48 \\ \hline \end{array}$

4. $\begin{array}{r} 475 \\ \times \quad 57 \\ \hline \end{array}$

5. $\begin{array}{r} 558 \\ \times \quad 64 \\ \hline \end{array}$

6. $\begin{array}{r} 398 \\ \times \quad 59 \\ \hline \end{array}$

7. $\begin{array}{r} 663 \\ \times \quad 49 \\ \hline \end{array}$

8. $\begin{array}{r} 583 \\ \times \quad 37 \\ \hline \end{array}$

9. $\begin{array}{r} 568 \\ \times \quad 68 \\ \hline \end{array}$

10. $\begin{array}{r} 438 \\ \times \quad 66 \\ \hline \end{array}$

11. $\begin{array}{r} 643 \\ \times \quad 57 \\ \hline \end{array}$

12. $\begin{array}{r} 592 \\ \times \quad 57 \\ \hline \end{array}$

Name: **Lesson 3-8 Multiplying two and three digit numbers**

1. $\begin{array}{r} 654 \\ \times\ 74 \\ \hline \end{array}$

2. $\begin{array}{r} 757 \\ \times\ 48 \\ \hline \end{array}$

3. $\begin{array}{r} 723 \\ \times\ 52 \\ \hline \end{array}$

4. $\begin{array}{r} 536 \\ \times\ 69 \\ \hline \end{array}$

5. $\begin{array}{r} 468 \\ \times\ 73 \\ \hline \end{array}$

6. $\begin{array}{r} 465 \\ \times\ 74 \\ \hline \end{array}$

7. $\begin{array}{r} 437 \\ \times\ 72 \\ \hline \end{array}$

8. $\begin{array}{r} 634 \\ \times\ 69 \\ \hline \end{array}$

9. $\begin{array}{r} 586 \\ \times\ 65 \\ \hline \end{array}$

10. $\begin{array}{r} 648 \\ \times\ 57 \\ \hline \end{array}$

11. $\begin{array}{r} 499 \\ \times\ 69 \\ \hline \end{array}$

12. $\begin{array}{r} 647 \\ \times\ 57 \\ \hline \end{array}$

Name: **Lesson 3-9 Multiplying two and three digit numbers**

1. $\begin{array}{r} 587 \\ \times \quad 63 \\ \hline \end{array}$

2. $\begin{array}{r} 577 \\ \times \quad 56 \\ \hline \end{array}$

3. $\begin{array}{r} 584 \\ \times \quad 67 \\ \hline \end{array}$

4. $\begin{array}{r} 833 \\ \times \quad 56 \\ \hline \end{array}$

5. $\begin{array}{r} 726 \\ \times \quad 48 \\ \hline \end{array}$

6. $\begin{array}{r} 746 \\ \times \quad 54 \\ \hline \end{array}$

7. $\begin{array}{r} 488 \\ \times \quad 75 \\ \hline \end{array}$

8. $\begin{array}{r} 657 \\ \times \quad 69 \\ \hline \end{array}$

9. $\begin{array}{r} 495 \\ \times \quad 77 \\ \hline \end{array}$

10. $\begin{array}{r} 597 \\ \times \quad 66 \\ \hline \end{array}$

11. $\begin{array}{r} 845 \\ \times \quad 78 \\ \hline \end{array}$

12. $\begin{array}{r} 834 \\ \times \quad 85 \\ \hline \end{array}$

Name: **Lesson 3-10 Multiplying two and three digit numbers**

---

1. $\begin{array}{r} 467 \\ \times \quad 73 \\ \hline \end{array}$

2. $\begin{array}{r} 453 \\ \times \quad 79 \\ \hline \end{array}$

3. $\begin{array}{r} 514 \\ \times \quad 88 \\ \hline \end{array}$

4. $\begin{array}{r} 656 \\ \times \quad 58 \\ \hline \end{array}$

5. $\begin{array}{r} 537 \\ \times \quad 84 \\ \hline \end{array}$

6. $\begin{array}{r} 636 \\ \times \quad 75 \\ \hline \end{array}$

7. $\begin{array}{r} 789 \\ \times \quad 84 \\ \hline \end{array}$

8. $\begin{array}{r} 756 \\ \times \quad 78 \\ \hline \end{array}$

9. $\begin{array}{r} 941 \\ \times \quad 92 \\ \hline \end{array}$

10. $\begin{array}{r} 489 \\ \times \quad 96 \\ \hline \end{array}$

11. $\begin{array}{r} 649 \\ \times \quad 93 \\ \hline \end{array}$

12. $\begin{array}{r} 878 \\ \times \quad 97 \\ \hline \end{array}$

Name: **Lesson 4-1 Dividing by one digit numbers with remainders**

1. $8\overline{)322}$
2. $4\overline{)311}$
3. $5\overline{)432}$
4. $7\overline{)472}$
5. $6\overline{)275}$
6. $7\overline{)496}$
7. $6\overline{)590}$
8. $4\overline{)311}$
9. $7\overline{)456}$
10. $9\overline{)245}$
11. $4\overline{)337}$
12. $6\overline{)298}$
13. $3\overline{)404}$
14. $6\overline{)386}$
15. $9\overline{)259}$
16. $4\overline{)365}$

Name: **Lesson 4-2 Dividing by one digit numbers with remainders**

1. $6\overline{)314}$ 2. $8\overline{)425}$ 3. $6\overline{)457}$ 4. $4\overline{)405}$

5. $7\overline{)562}$ 6. $9\overline{)823}$ 7. $8\overline{)538}$ 8. $8\overline{)510}$

9. $9\overline{)359}$ 10. $7\overline{)446}$ 11. $7\overline{)447}$ 12. $6\overline{)493}$

13. $3\overline{)421}$ 14. $7\overline{)512}$ 15. $9\overline{)348}$ 16. $8\overline{)466}$

Name: **Lesson 4-3 Dividing by one digit numbers with remainders**

1. $8\overline{)438}$	2. $7\overline{)424}$	3. $9\overline{)565}$	4. $4\overline{)418}$
5. $6\overline{)536}$	6. $6\overline{)404}$	7. $7\overline{)318}$	8. $6\overline{)543}$
9. $4\overline{)333}$	10. $3\overline{)299}$	11. $3\overline{)220}$	12. $7\overline{)387}$
13. $5\overline{)492}$	14. $7\overline{)512}$	15. $8\overline{)503}$	16. $9\overline{)564}$

Name:

Lesson 4-4 Dividing by one digit numbers with remainders

1. $7\overline{)391}$ 2. $5\overline{)384}$ 3. $7\overline{)355}$ 4. $4\overline{)257}$

5. $5\overline{)474}$ 6. $3\overline{)500}$ 7. $6\overline{)479}$ 8. $8\overline{)404}$

9. $6\overline{)555}$ 10. $7\overline{)443}$ 11. $6\overline{)292}$ 12. $6\overline{)301}$

13. $9\overline{)405}$ 14. $4\overline{)398}$ 15. $8\overline{)551}$ 16. $9\overline{)563}$

Name: **Lesson 4-5 Dividing by one digit numbers with remainders**

1. $3\overline{)290}$	2. $4\overline{)219}$	3. $7\overline{)377}$	4. $7\overline{)526}$
5. $4\overline{)375}$	6. $7\overline{)520}$	7. $4\overline{)321}$	8. $6\overline{)555}$
9. $7\overline{)404}$	10. $9\overline{)499}$	11. $6\overline{)579}$	12. $9\overline{)470}$
13. $8\overline{)565}$	14. $3\overline{)436}$	15. $3\overline{)488}$	16. $7\overline{)383}$

Name: **Lesson 4-6 Dividing by one digit numbers with remainders**

1. $5\overline{)366}$	2. $3\overline{)412}$	3. $8\overline{)530}$	4. $3\overline{)423}$
5. $6\overline{)632}$	6. $7\overline{)556}$	7. $7\overline{)499}$	8. $7\overline{)667}$
9. $7\overline{)554}$	10. $8\overline{)333}$	11. $9\overline{)222}$	12. $4\overline{)461}$
13. $8\overline{)447}$	14. $9\overline{)664}$	15. $9\overline{)619}$	16. $7\overline{)600}$

Name: **Lesson 4-7 Dividing by one digit numbers with remainders**

1. $6\overline{)511}$ 2. $5\overline{)461}$ 3. $4\overline{)537}$ 4. $7\overline{)613}$

5. $7\overline{)603}$ 6. $9\overline{)452}$ 7. $7\overline{)466}$ 8. $5\overline{)544}$

9. $8\overline{)457}$ 10. $7\overline{)523}$ 11. $8\overline{)577}$ 12. $9\overline{)632}$

13. $9\overline{)699}$ 14. $8\overline{)670}$ 15. $4\overline{)330}$ 16. $8\overline{)614}$

Name: **Lesson 4-8 Dividing by one digit numbers with remainders**

1. $7\overline{)732}$

2. $7\overline{)662}$

3. $8\overline{)631}$

4. $4\overline{)789}$

5. $5\overline{)638}$

6. $9\overline{)521}$

7. $7\overline{)300}$

8. $9\overline{)599}$

9. $9\overline{)619}$

10. $6\overline{)431}$

11. $4\overline{)411}$

12. $8\overline{)631}$

13. $8\overline{)474}$

14. $9\overline{)556}$

15. $6\overline{)509}$

16. $7\overline{)551}$

Name:

Lesson 4-9 Dividing by one digit numbers with remainders

1. $3\overline{)700}$ 2. $8\overline{)711}$ 3. $9\overline{)604}$ 4. $6\overline{)665}$

5. $9\overline{)446}$ 6. $7\overline{)624}$ 7. $6\overline{)537}$ 8. $9\overline{)573}$

9. $7\overline{)437}$ 10. $9\overline{)453}$ 11. $4\overline{)639}$ 12. $6\overline{)333}$

13. $6\overline{)752}$ 14. $8\overline{)729}$ 15. $9\overline{)771}$ 16. $7\overline{)548}$

Name:

Lesson 4-10 Dividing by one digit numbers with remainders

1. $9\overline{)821}$	2. $3\overline{)652}$	3. $8\overline{)634}$	4. $7\overline{)579}$
5. $8\overline{)734}$	6. $8\overline{)700}$	7. $5\overline{)766}$	8. $9\overline{)732}$
9. $6\overline{)773}$	10. $7\overline{)646}$	11. $9\overline{)541}$	12. $8\overline{)623}$
13. $5\overline{)622}$	14. $9\overline{)746}$	15. $3\overline{)698}$	16. $4\overline{)571}$

Name: **Lesson 5-1 Multiplying three digit numbers**

1. $\begin{array}{r} 332 \\ \times\ 416 \\ \hline \end{array}$

2. $\begin{array}{r} 273 \\ \times\ 457 \\ \hline \end{array}$

3. $\begin{array}{r} 445 \\ \times\ 379 \\ \hline \end{array}$

4. $\begin{array}{r} 428 \\ \times\ 439 \\ \hline \end{array}$

5. $\begin{array}{r} 529 \\ \times\ 375 \\ \hline \end{array}$

6. $\begin{array}{r} 362 \\ \times\ 348 \\ \hline \end{array}$

7. $\begin{array}{r} 289 \\ \times\ 598 \\ \hline \end{array}$

8. $\begin{array}{r} 374 \\ \times\ 478 \\ \hline \end{array}$

9. $\begin{array}{r} 469 \\ \times\ 395 \\ \hline \end{array}$

Name: **Lesson 5-2 Multiplying three digit numbers**

1. $\begin{array}{r} 349 \\ \times\ 589 \\ \hline \end{array}$

2. $\begin{array}{r} 492 \\ \times\ 456 \\ \hline \end{array}$

3. $\begin{array}{r} 546 \\ \times\ 387 \\ \hline \end{array}$

4. $\begin{array}{r} 553 \\ \times\ 355 \\ \hline \end{array}$

5. $\begin{array}{r} 517 \\ \times\ 636 \\ \hline \end{array}$

6. $\begin{array}{r} 657 \\ \times\ 482 \\ \hline \end{array}$

7. $\begin{array}{r} 547 \\ \times\ 543 \\ \hline \end{array}$

8. $\begin{array}{r} 579 \\ \times\ 434 \\ \hline \end{array}$

9. $\begin{array}{r} 349 \\ \times\ 529 \\ \hline \end{array}$

Name: **Lesson 5-3 Multiplying three digit numbers**

1. $\begin{array}{r} 442 \\ \times\ 326 \\ \hline \end{array}$

2. $\begin{array}{r} 458 \\ \times\ 394 \\ \hline \end{array}$

3. $\begin{array}{r} 492 \\ \times\ 357 \\ \hline \end{array}$

4. $\begin{array}{r} 574 \\ \times\ 468 \\ \hline \end{array}$

5. $\begin{array}{r} 563 \\ \times\ 536 \\ \hline \end{array}$

6. $\begin{array}{r} 531 \\ \times\ 642 \\ \hline \end{array}$

7. $\begin{array}{r} 641 \\ \times\ 515 \\ \hline \end{array}$

8. $\begin{array}{r} 389 \\ \times\ 398 \\ \hline \end{array}$

9. $\begin{array}{r} 365 \\ \times\ 497 \\ \hline \end{array}$

Name: **Lesson 5-4 Multiplying three digit numbers**

1.
$$\begin{array}{r} 449 \\ \times\ 644 \\ \hline \end{array}$$

2.
$$\begin{array}{r} 409 \\ \times\ 499 \\ \hline \end{array}$$

3.
$$\begin{array}{r} 573 \\ \times\ 649 \\ \hline \end{array}$$

4.
$$\begin{array}{r} 673 \\ \times\ 455 \\ \hline \end{array}$$

5.
$$\begin{array}{r} 557 \\ \times\ 587 \\ \hline \end{array}$$

6.
$$\begin{array}{r} 624 \\ \times\ 525 \\ \hline \end{array}$$

7.
$$\begin{array}{r} 538 \\ \times\ 432 \\ \hline \end{array}$$

8.
$$\begin{array}{r} 645 \\ \times\ 564 \\ \hline \end{array}$$

9.
$$\begin{array}{r} 573 \\ \times\ 697 \\ \hline \end{array}$$

Name: **Lesson 5-5 Multiplying three digit numbers**

1.
$$\begin{array}{r} 4\,2\,9 \\ \times\ 5\,4\,3 \\ \hline \end{array}$$

2.
$$\begin{array}{r} 3\,5\,6 \\ \times\ 6\,2\,7 \\ \hline \end{array}$$

3.
$$\begin{array}{r} 4\,8\,7 \\ \times\ 3\,7\,9 \\ \hline \end{array}$$

4.
$$\begin{array}{r} 4\,8\,9 \\ \times\ 4\,7\,5 \\ \hline \end{array}$$

5.
$$\begin{array}{r} 5\,8\,9 \\ \times\ 5\,1\,9 \\ \hline \end{array}$$

6.
$$\begin{array}{r} 5\,3\,7 \\ \times\ 4\,4\,8 \\ \hline \end{array}$$

7.
$$\begin{array}{r} 4\,6\,7 \\ \times\ 6\,8\,4 \\ \hline \end{array}$$

8.
$$\begin{array}{r} 6\,8\,5 \\ \times\ 6\,7\,6 \\ \hline \end{array}$$

9.
$$\begin{array}{r} 3\,8\,7 \\ \times\ 3\,9\,9 \\ \hline \end{array}$$

Name: **Lesson 5-6 Multiplying three digit numbers**

1. $\begin{array}{r} 351 \\ \times\ 645 \\ \hline \end{array}$

2. $\begin{array}{r} 428 \\ \times\ 648 \\ \hline \end{array}$

3. $\begin{array}{r} 462 \\ \times\ 532 \\ \hline \end{array}$

4. $\begin{array}{r} 431 \\ \times\ 688 \\ \hline \end{array}$

5. $\begin{array}{r} 674 \\ \times\ 597 \\ \hline \end{array}$

6. $\begin{array}{r} 667 \\ \times\ 344 \\ \hline \end{array}$

7. $\begin{array}{r} 507 \\ \times\ 417 \\ \hline \end{array}$

8. $\begin{array}{r} 645 \\ \times\ 352 \\ \hline \end{array}$

9. $\begin{array}{r} 424 \\ \times\ 535 \\ \hline \end{array}$

Name: **Lesson 5-7 Multiplying three digit numbers**

1. $\begin{array}{r} 374 \\ \times\ 575 \\ \hline \end{array}$

2. $\begin{array}{r} 587 \\ \times\ 696 \\ \hline \end{array}$

3. $\begin{array}{r} 597 \\ \times\ 593 \\ \hline \end{array}$

4. $\begin{array}{r} 684 \\ \times\ 784 \\ \hline \end{array}$

5. $\begin{array}{r} 758 \\ \times\ 493 \\ \hline \end{array}$

6. $\begin{array}{r} 627 \\ \times\ 733 \\ \hline \end{array}$

7. $\begin{array}{r} 776 \\ \times\ 376 \\ \hline \end{array}$

8. $\begin{array}{r} 591 \\ \times\ 724 \\ \hline \end{array}$

9. $\begin{array}{r} 468 \\ \times\ 468 \\ \hline \end{array}$

Name: **Lesson 5-8 Multiplying three digit numbers**

1.
$$\begin{array}{r} 456 \\ \times\ 579 \\ \hline \end{array}$$

2.
$$\begin{array}{r} 497 \\ \times\ 486 \\ \hline \end{array}$$

3.
$$\begin{array}{r} 452 \\ \times\ 499 \\ \hline \end{array}$$

4.
$$\begin{array}{r} 658 \\ \times\ 452 \\ \hline \end{array}$$

5.
$$\begin{array}{r} 753 \\ \times\ 375 \\ \hline \end{array}$$

6.
$$\begin{array}{r} 683 \\ \times\ 615 \\ \hline \end{array}$$

7.
$$\begin{array}{r} 741 \\ \times\ 489 \\ \hline \end{array}$$

8.
$$\begin{array}{r} 636 \\ \times\ 531 \\ \hline \end{array}$$

9.
$$\begin{array}{r} 493 \\ \times\ 793 \\ \hline \end{array}$$

Name: **Lesson 5-9 Multiplying three digit numbers**

1.
$$\begin{array}{r} 484 \\ \times\ 776 \\ \hline \end{array}$$

2.
$$\begin{array}{r} 557 \\ \times\ 586 \\ \hline \end{array}$$

3.
$$\begin{array}{r} 488 \\ \times\ 568 \\ \hline \end{array}$$

4.
$$\begin{array}{r} 658 \\ \times\ 547 \\ \hline \end{array}$$

5.
$$\begin{array}{r} 717 \\ \times\ 526 \\ \hline \end{array}$$

6.
$$\begin{array}{r} 674 \\ \times\ 698 \\ \hline \end{array}$$

7.
$$\begin{array}{r} 773 \\ \times\ 577 \\ \hline \end{array}$$

8.
$$\begin{array}{r} 743 \\ \times\ 357 \\ \hline \end{array}$$

9.
$$\begin{array}{r} 575 \\ \times\ 683 \\ \hline \end{array}$$

Name: **Lesson 5-10 Multiplying three digit numbers**

1. $\begin{array}{r} 589 \\ \times\ 677 \\ \hline \end{array}$

2. $\begin{array}{r} 565 \\ \times\ 487 \\ \hline \end{array}$

3. $\begin{array}{r} 465 \\ \times\ 564 \\ \hline \end{array}$

4. $\begin{array}{r} 777 \\ \times\ 454 \\ \hline \end{array}$

5. $\begin{array}{r} 783 \\ \times\ 409 \\ \hline \end{array}$

6. $\begin{array}{r} 784 \\ \times\ 678 \\ \hline \end{array}$

7. $\begin{array}{r} 687 \\ \times\ 642 \\ \hline \end{array}$

8. $\begin{array}{r} 834 \\ \times\ 513 \\ \hline \end{array}$

9. $\begin{array}{r} 775 \\ \times\ 567 \\ \hline \end{array}$

Name:

Lesson 6-1 Dividing by two digit numbers with remainders

1. $32\overline{\smash{)}654}$
2. $37\overline{\smash{)}572}$
3. $31\overline{\smash{)}576}$
4. $34\overline{\smash{)}545}$
5. $25\overline{\smash{)}498}$
6. $46\overline{\smash{)}661}$
7. $27\overline{\smash{)}468}$
8. $26\overline{\smash{)}488}$
9. $17\overline{\smash{)}351}$
10. $38\overline{\smash{)}595}$
11. $39\overline{\smash{)}549}$
12. $19\overline{\smash{)}461}$
13. $19\overline{\smash{)}565}$
14. $23\overline{\smash{)}512}$
15. $17\overline{\smash{)}478}$
16. $37\overline{\smash{)}475}$

Name: **Lesson 6-2 Dividing by two digit numbers with remainders**

1. $43\overline{)592}$ 2. $15\overline{)778}$ 3. $16\overline{)653}$ 4. $27\overline{)774}$

5. $31\overline{)624}$ 6. $27\overline{)660}$ 7. $28\overline{)598}$ 8. $35\overline{)688}$

9. $37\overline{)505}$ 10. $29\overline{)655}$ 11. $31\overline{)600}$ 12. $17\overline{)611}$

13. $42\overline{)633}$ 14. $26\overline{)735}$ 15. $29\overline{)577}$ 16. $21\overline{)594}$

Name: **Lesson 6-3 Dividing by two digit numbers with remainders**

1. $34\overline{)753}$	2. $24\overline{)679}$	3. $35\overline{)766}$	4. $31\overline{)735}$
5. $41\overline{)659}$	6. $16\overline{)770}$	7. $17\overline{)635}$	8. $18\overline{)639}$
9. $15\overline{)531}$	10. $26\overline{)622}$	11. $13\overline{)757}$	12. $19\overline{)686}$
13. $16\overline{)790}$	14. $17\overline{)696}$	15. $37\overline{)564}$	16. $26\overline{)547}$

Name:

Lesson 6-4 Dividing by two digit numbers with remainders

1. $41\,\overline{)831}$
2. $34\,\overline{)767}$
3. $42\,\overline{)589}$
4. $63\,\overline{)844}$
5. $24\,\overline{)767}$
6. $19\,\overline{)864}$
7. $16\,\overline{)838}$
8. $23\,\overline{)715}$
9. $17\,\overline{)662}$
10. $53\,\overline{)743}$
11. $56\,\overline{)726}$
12. $12\,\overline{)767}$
13. $35\,\overline{)786}$
14. $17\,\overline{)678}$
15. $12\,\overline{)599}$
16. $21\,\overline{)620}$

Name: **Lesson 6-5 Dividing by two digit numbers with remainders**

1. $15\overline{)632}$ 2. $14\overline{)439}$ 3. $35\overline{)599}$ 4. $24\overline{)754}$

5. $20\overline{)765}$ 6. $27\overline{)566}$ 7. $11\overline{)497}$ 8. $13\overline{)633}$

9. $17\overline{)776}$ 10. $43\overline{)736}$ 11. $24\overline{)783}$ 12. $28\overline{)675}$

13. $35\overline{)647}$ 14. $39\overline{)681}$ 15. $43\overline{)595}$ 16. $15\overline{)714}$

Name: **Lesson 6-6 Dividing by two digit numbers with remainders**

---

1. $22\overline{)641}$ 2. $36\overline{)772}$ 3. $52\overline{)714}$ 4. $54\overline{)781}$

5. $45\overline{)589}$ 6. $23\overline{)654}$ 7. $21\overline{)582}$ 8. $33\overline{)655}$

9. $12\overline{)785}$ 10. $12\overline{)789}$ 11. $16\overline{)627}$ 12. $11\overline{)798}$

13. $37\overline{)644}$ 14. $31\overline{)780}$ 15. $39\overline{)737}$ 16. $29\overline{)692}$

Name:

Lesson 6-7 Dividing by two digit numbers with remainders

1. $41\overline{)891}$	2. $47\overline{)682}$	3. $26\overline{)641}$	4. $45\overline{)698}$
5. $12\overline{)475}$	6. $24\overline{)793}$	7. $25\overline{)757}$	8. $52\overline{)781}$
9. $34\overline{)743}$	10. $13\overline{)804}$	11. $19\overline{)864}$	12. $13\overline{)852}$
13. $16\overline{)813}$	14. $27\overline{)710}$	15. $33\overline{)793}$	16. $25\overline{)785}$

Name: **Lesson 6-8 Dividing by two digit numbers with remainders**

1. $25\overline{)555}$

2. $34\overline{)522}$

3. $21\overline{)697}$

4. $45\overline{)739}$

5. $11\overline{)790}$

6. $23\overline{)817}$

7. $33\overline{)546}$

8. $32\overline{)863}$

9. $41\overline{)812}$

10. $16\overline{)886}$

11. $16\overline{)887}$

12. $24\overline{)650}$

13. $63\overline{)836}$

14. $32\overline{)737}$

15. $27\overline{)705}$

16. $16\overline{)707}$

Name: **Lesson 6-9 Dividing by two digit numbers with remainders**

1. $24\,\overline{)745}$

2. $29\,\overline{)964}$

3. $68\,\overline{)840}$

4. $32\,\overline{)832}$

5. $29\,\overline{)842}$

6. $45\,\overline{)747}$

7. $45\,\overline{)703}$

8. $17\,\overline{)713}$

9. $37\,\overline{)779}$

10. $34\,\overline{)518}$

11. $25\,\overline{)942}$

12. $41\,\overline{)622}$

13. $18\,\overline{)862}$

14. $13\,\overline{)999}$

15. $14\,\overline{)493}$

16. $28\,\overline{)961}$

Name: **Lesson 6-10 Dividing by two digit numbers with remainders**

1. $33\overline{)753}$
2. $17\overline{)982}$
3. $12\overline{)866}$
4. $54\overline{)953}$
5. $26\overline{)756}$
6. $28\overline{)795}$
7. $17\overline{)865}$
8. $19\overline{)802}$
9. $31\overline{)610}$
10. $32\overline{)840}$
11. $28\overline{)580}$
12. $37\overline{)789}$
13. $18\overline{)834}$
14. $11\overline{)652}$
15. $49\overline{)722}$
16. $25\overline{)855}$

Name: **Lesson 7-1 Multiplying three digit numbers**

1. $\begin{array}{r} 412 \\ \times\ 497 \\ \hline \end{array}$

2. $\begin{array}{r} 439 \\ \times\ 572 \\ \hline \end{array}$

3. $\begin{array}{r} 524 \\ \times\ 479 \\ \hline \end{array}$

4. $\begin{array}{r} 571 \\ \times\ 589 \\ \hline \end{array}$

5. $\begin{array}{r} 681 \\ \times\ 476 \\ \hline \end{array}$

6. $\begin{array}{r} 297 \\ \times\ 548 \\ \hline \end{array}$

7. $\begin{array}{r} 728 \\ \times\ 676 \\ \hline \end{array}$

8. $\begin{array}{r} 683 \\ \times\ 649 \\ \hline \end{array}$

9. $\begin{array}{r} 649 \\ \times\ 567 \\ \hline \end{array}$

Name: **Lesson 7-2 Multiplying three digit numbers**

1.
$$\begin{array}{r} 577 \\ \times\ 632 \\ \hline \end{array}$$

2.
$$\begin{array}{r} 357 \\ \times\ 542 \\ \hline \end{array}$$

3.
$$\begin{array}{r} 675 \\ \times\ 294 \\ \hline \end{array}$$

4.
$$\begin{array}{r} 546 \\ \times\ 759 \\ \hline \end{array}$$

5.
$$\begin{array}{r} 641 \\ \times\ 682 \\ \hline \end{array}$$

6.
$$\begin{array}{r} 593 \\ \times\ 745 \\ \hline \end{array}$$

7.
$$\begin{array}{r} 845 \\ \times\ 678 \\ \hline \end{array}$$

8.
$$\begin{array}{r} 734 \\ \times\ 457 \\ \hline \end{array}$$

9.
$$\begin{array}{r} 842 \\ \times\ 297 \\ \hline \end{array}$$

1. $\begin{array}{r} 424 \\ \times\ 478 \\ \hline \end{array}$

2. $\begin{array}{r} 682 \\ \times\ 387 \\ \hline \end{array}$

3. $\begin{array}{r} 553 \\ \times\ 629 \\ \hline \end{array}$

4. $\begin{array}{r} 653 \\ \times\ 365 \\ \hline \end{array}$

5. $\begin{array}{r} 561 \\ \times\ 396 \\ \hline \end{array}$

6. $\begin{array}{r} 578 \\ \times\ 724 \\ \hline \end{array}$

7. $\begin{array}{r} 742 \\ \times\ 687 \\ \hline \end{array}$

8. $\begin{array}{r} 498 \\ \times\ 765 \\ \hline \end{array}$

9. $\begin{array}{r} 625 \\ \times\ 794 \\ \hline \end{array}$

Name: **Lesson 7-4 Multiplying three digit numbers**

1.
$$\begin{array}{r} 4\,2\,9 \\ \times\ 7\,8\,4 \\ \hline \end{array}$$

2.
$$\begin{array}{r} 4\,7\,5 \\ \times\ 4\,8\,8 \\ \hline \end{array}$$

3.
$$\begin{array}{r} 4\,5\,8 \\ \times\ 6\,5\,2 \\ \hline \end{array}$$

4.
$$\begin{array}{r} 8\,9\,2 \\ \times\ 3\,1\,5 \\ \hline \end{array}$$

5.
$$\begin{array}{r} 9\,6\,4 \\ \times\ 4\,7\,9 \\ \hline \end{array}$$

6.
$$\begin{array}{r} 9\,3\,5 \\ \times\ 6\,7\,8 \\ \hline \end{array}$$

7.
$$\begin{array}{r} 5\,3\,8 \\ \times\ 6\,7\,6 \\ \hline \end{array}$$

8.
$$\begin{array}{r} 8\,5\,2 \\ \times\ 4\,7\,2 \\ \hline \end{array}$$

9.
$$\begin{array}{r} 4\,9\,9 \\ \times\ 3\,6\,6 \\ \hline \end{array}$$

Name: **Lesson 7-5 Multiplying three digit numbers**

1.
$$\begin{array}{r} 392 \\ \times\ 478 \\ \hline \end{array}$$

2.
$$\begin{array}{r} 489 \\ \times\ 564 \\ \hline \end{array}$$

3.
$$\begin{array}{r} 742 \\ \times\ 648 \\ \hline \end{array}$$

4.
$$\begin{array}{r} 756 \\ \times\ 843 \\ \hline \end{array}$$

5.
$$\begin{array}{r} 667 \\ \times\ 733 \\ \hline \end{array}$$

6.
$$\begin{array}{r} 537 \\ \times\ 559 \\ \hline \end{array}$$

7.
$$\begin{array}{r} 556 \\ \times\ 696 \\ \hline \end{array}$$

8.
$$\begin{array}{r} 598 \\ \times\ 346 \\ \hline \end{array}$$

9.
$$\begin{array}{r} 728 \\ \times\ 736 \\ \hline \end{array}$$

Name: Lesson 7-6 Multiplying three digit numbers

1. $\begin{array}{r} 256 \\ \times\ 715 \\ \hline \end{array}$

2. $\begin{array}{r} 456 \\ \times\ 789 \\ \hline \end{array}$

3. $\begin{array}{r} 575 \\ \times\ 636 \\ \hline \end{array}$

4. $\begin{array}{r} 563 \\ \times\ 946 \\ \hline \end{array}$

5. $\begin{array}{r} 723 \\ \times\ 478 \\ \hline \end{array}$

6. $\begin{array}{r} 696 \\ \times\ 799 \\ \hline \end{array}$

7. $\begin{array}{r} 835 \\ \times\ 875 \\ \hline \end{array}$

8. $\begin{array}{r} 592 \\ \times\ 282 \\ \hline \end{array}$

9. $\begin{array}{r} 537 \\ \times\ 655 \\ \hline \end{array}$

Name: **Lesson 7-7 Multiplying three digit numbers**

1.
$$\begin{array}{r} 466 \\ \times\ 674 \\ \hline \end{array}$$

2.
$$\begin{array}{r} 743 \\ \times\ 365 \\ \hline \end{array}$$

3.
$$\begin{array}{r} 894 \\ \times\ 745 \\ \hline \end{array}$$

4.
$$\begin{array}{r} 695 \\ \times\ 658 \\ \hline \end{array}$$

5.
$$\begin{array}{r} 459 \\ \times\ 424 \\ \hline \end{array}$$

6.
$$\begin{array}{r} 783 \\ \times\ 594 \\ \hline \end{array}$$

7.
$$\begin{array}{r} 758 \\ \times\ 549 \\ \hline \end{array}$$

8.
$$\begin{array}{r} 647 \\ \times\ 683 \\ \hline \end{array}$$

9.
$$\begin{array}{r} 672 \\ \times\ 521 \\ \hline \end{array}$$

Name: **Lesson 7-8 Multiplying three digit numbers**

1. $\begin{array}{r} 324 \\ \times\ 487 \\ \hline \end{array}$

2. $\begin{array}{r} 687 \\ \times\ 758 \\ \hline \end{array}$

3. $\begin{array}{r} 346 \\ \times\ 382 \\ \hline \end{array}$

4. $\begin{array}{r} 458 \\ \times\ 976 \\ \hline \end{array}$

5. $\begin{array}{r} 865 \\ \times\ 973 \\ \hline \end{array}$

6. $\begin{array}{r} 679 \\ \times\ 592 \\ \hline \end{array}$

7. $\begin{array}{r} 765 \\ \times\ 876 \\ \hline \end{array}$

8. $\begin{array}{r} 445 \\ \times\ 871 \\ \hline \end{array}$

9. $\begin{array}{r} 585 \\ \times\ 864 \\ \hline \end{array}$

Name: **Lesson 7-9 Multiplying three digit numbers**

1. $\begin{array}{r} 726 \\ \times\ 973 \\ \hline \end{array}$

2. $\begin{array}{r} 734 \\ \times\ 568 \\ \hline \end{array}$

3. $\begin{array}{r} 902 \\ \times\ 324 \\ \hline \end{array}$

4. $\begin{array}{r} 458 \\ \times\ 912 \\ \hline \end{array}$

5. $\begin{array}{r} 589 \\ \times\ 635 \\ \hline \end{array}$

6. $\begin{array}{r} 664 \\ \times\ 727 \\ \hline \end{array}$

7. $\begin{array}{r} 776 \\ \times\ 878 \\ \hline \end{array}$

8. $\begin{array}{r} 599 \\ \times\ 399 \\ \hline \end{array}$

9. $\begin{array}{r} 879 \\ \times\ 847 \\ \hline \end{array}$

Name: **Lesson 7-10 Multiplying three digit numbers**

1.
$$\begin{array}{r} 849 \\ \times\ 634 \\ \hline \end{array}$$

2.
$$\begin{array}{r} 367 \\ \times\ 498 \\ \hline \end{array}$$

3.
$$\begin{array}{r} 645 \\ \times\ 862 \\ \hline \end{array}$$

4.
$$\begin{array}{r} 964 \\ \times\ 969 \\ \hline \end{array}$$

5.
$$\begin{array}{r} 609 \\ \times\ 369 \\ \hline \end{array}$$

6.
$$\begin{array}{r} 971 \\ \times\ 538 \\ \hline \end{array}$$

7.
$$\begin{array}{r} 875 \\ \times\ 385 \\ \hline \end{array}$$

8.
$$\begin{array}{r} 517 \\ \times\ 628 \\ \hline \end{array}$$

9.
$$\begin{array}{r} 556 \\ \times\ 975 \\ \hline \end{array}$$

Name: **Lesson 8-1 Dividing by two digit numbers with remainders**

1. $35\overline{)672}$	2. $64\overline{)861}$	3. $12\overline{)523}$	4. $27\overline{)645}$
5. $11\overline{)239}$	6. $26\overline{)790}$	7. $56\overline{)773}$	8. $25\overline{)992}$
9. $42\overline{)434}$	10. $29\overline{)389}$	11. $54\overline{)632}$	12. $27\overline{)784}$
13. $13\overline{)357}$	14. $31\overline{)512}$	15. $42\overline{)845}$	16. $32\overline{)647}$

Name: **Lesson 8-2 Dividing by two digit numbers with remainders**

1. $58\overline{)881}$
2. $37\overline{)650}$
3. $12\overline{)495}$
4. $22\overline{)784}$
5. $41\overline{)452}$
6. $24\overline{)779}$
7. $54\overline{)680}$
8. $49\overline{)941}$
9. $25\overline{)707}$
10. $14\overline{)997}$
11. $52\overline{)587}$
12. $63\overline{)633}$
13. $13\overline{)919}$
14. $43\overline{)584}$
15. $23\overline{)771}$
16. $17\overline{)872}$

Name: **Lesson 8-3 Dividing by two digit numbers with remainders**

---

1. $29\overline{)471}$	2. $37\overline{)750}$	3. $42\overline{)567}$	4. $15\overline{)832}$
5. $36\overline{)752}$	6. $28\overline{)960}$	7. $73\overline{)810}$	8. $56\overline{)566}$
9. $17\overline{)976}$	10. $21\overline{)530}$	11. $22\overline{)795}$	12. $24\overline{)731}$
13. $43\overline{)592}$	14. $17\overline{)777}$	15. $15\overline{)836}$	16. $48\overline{)899}$

Name: **Lesson 8-4 Dividing by two digit numbers with remainders**

---

1. $17\overline{)742}$ 2. $26\overline{)875}$ 3. $47\overline{)892}$ 4. $14\overline{)594}$

5. $25\overline{)660}$ 6. $14\overline{)461}$ 7. $53\overline{)542}$ 8. $16\overline{)633}$

9. $38\overline{)761}$ 10. $39\overline{)723}$ 11. $27\overline{)725}$ 12. $23\overline{)743}$

13. $13\overline{)948}$ 14. $67\overline{)945}$ 15. $58\overline{)987}$ 16. $32\overline{)837}$

Name: **Lesson 8-5 Dividing by two digit numbers with remainders**

1. $42\overline{)734}$

2. $34\overline{)561}$

3. $47\overline{)680}$

4. $31\overline{)955}$

5. $57\overline{)892}$

6. $19\overline{)556}$

7. $22\overline{)774}$

8. $16\overline{)654}$

9. $39\overline{)740}$

10. $44\overline{)649}$

11. $33\overline{)748}$

12. $22\overline{)784}$

13. $27\overline{)635}$

14. $29\overline{)888}$

15. $27\overline{)976}$

16. $42\overline{)972}$

Name: **Lesson 8-6 Dividing by two digit numbers with remainders**

1. $24\overline{)467}$
2. $54\overline{)894}$
3. $31\overline{)999}$
4. $14\overline{)551}$
5. $55\overline{)844}$
6. $22\overline{)671}$
7. $25\overline{)702}$
8. $33\overline{)922}$
9. $17\overline{)532}$
10. $37\overline{)956}$
11. $47\overline{)818}$
12. $26\overline{)827}$
13. $39\overline{)998}$
14. $41\overline{)749}$
15. $27\overline{)954}$
16. $65\overline{)997}$

Name: **Lesson 8-7 Dividing by two digit numbers with remainders**

1. $44\overline{)891}$	2. $41\overline{)785}$	3. $63\overline{)651}$	4. $26\overline{)930}$
5. $31\overline{)838}$	6. $26\overline{)587}$	7. $15\overline{)520}$	8. $46\overline{)871}$
9. $12\overline{)393}$	10. $13\overline{)636}$	11. $29\overline{)596}$	12. $35\overline{)952}$
13. $27\overline{)712}$	14. $16\overline{)904}$	15. $44\overline{)699}$	16. $11\overline{)478}$

Name:

Lesson 8-8 Dividing by two digit numbers with remainders

1. $34\overline{)555}$
2. $29\overline{)542}$
3. $21\overline{)983}$
4. $28\overline{)958}$
5. $28\overline{)709}$
6. $14\overline{)841}$
7. $32\overline{)940}$
8. $29\overline{)798}$
9. $39\overline{)526}$
10. $16\overline{)737}$
11. $37\overline{)760}$
12. $38\overline{)575}$
13. $57\overline{)683}$
14. $37\overline{)958}$
15. $46\overline{)825}$
16. $18\overline{)961}$

Name: **Lesson 8-9 Dividing by two digit numbers with remainders**

1. $42\overline{)753}$	2. $57\overline{)976}$	3. $48\overline{)840}$	4. $24\overline{)854}$
5. $26\overline{)671}$	6. $22\overline{)853}$	7. $19\overline{)580}$	8. $47\overline{)669}$
9. $38\overline{)555}$	10. $33\overline{)824}$	11. $23\overline{)652}$	12. $12\overline{)473}$
13. $13\overline{)772}$	14. $14\overline{)981}$	15. $32\overline{)875}$	16. $41\overline{)787}$

Name: **Lesson 8-10 Dividing by two digit numbers with remainders**

---

1. $17\overline{)482}$ 2. $39\overline{)872}$ 3. $17\overline{)488}$ 4. $54\overline{)974}$

5. $32\overline{)886}$ 6. $22\overline{)589}$ 7. $29\overline{)893}$ 8. $16\overline{)990}$

9. $13\overline{)310}$ 10. $31\overline{)820}$ 11. $31\overline{)850}$ 12. $37\overline{)705}$

13. $27\overline{)832}$ 14. $42\overline{)955}$ 15. $26\overline{)735}$ 16. $21\overline{)689}$

1. $13 + 5 - 7 =$
2. $6 \times 9 - 18 =$
3. $24 - 3 \times 6 =$
4. $28 \div 4 + 13 =$
5. $24 \div 3 + 8 =$
6. $(23 - 5) \times 3 =$
7. $5 \times (25 - 6) =$
8. $72 \div (23 - 11) =$
9. $32 + 3 \times 4 =$
10. $55 \div (4 + 7) =$
11. $36 \div 2 - 9 =$
12. $15 - 81 \div 9 =$
13. $(7 + 8) \times 4 =$
14. $3 \times (4 + 7) =$
15. $9 + 30 \div 5 =$
16. $44 - 17 - 21 =$
17. $54 \div 9 \times 6 =$
18. $8 \times 5 + 27 =$
19. $64 \div 8 \div 2 =$
20. $16 \times 4 \div 8 =$
21. $(24 + 12) \div 9 =$
22. $4 \times 5 \times 6 =$
23. $(54 \div 6) \times 9 =$
24. $(23 - 5) \div 6 =$

1. $17 - 9 + 8 =$
2. $40 \div 8 + 7 =$
3. $33 - 2 \times 9 =$
4. $5 \times 9 - 8 =$
5. $7 \times (11 - 3) =$
6. $84 \div (5 + 7) =$
7. $25 + 11 + 9 =$
8. $(27 - 13) \times 8 =$
9. $45 + 3 \times 9 =$
10. $5 \times (7 + 4) =$
11. $6 \times (16 + 7) =$
12. $37 - (64 \div 4) =$
13. $(7 + 8) \times 4 =$
14. $65 - 23 - 19 =$
15. $90 \div (15 - 9) =$
16. $42 \div 7 \times 8 =$
17. $56 \div 8 \times 7 =$
18. $4 \times 8 + 15 =$
19. $13 + 35 \div 7 =$
20. $55 - 18 + 23 =$
21. $(29 + 27) \div 7 =$
22. $(32 - 7) \div 5 =$
23. $99 \div 3 \div 11 =$
24. $4 \times 3 \times 9 =$

1. $28 + 8 - 7 =$
2. $5 \times 7 - 18 =$
3. $67 - 9 \times 6 =$
4. $72 \div 6 + 15 =$
5. $14 + 27 - 13 =$
6. $(54 - 38) \times 4 =$
7. $6 \times (21 - 7) =$
8. $60 \div (6 + 4) =$
9. $33 + 4 \times 8 =$
10. $(42 - 18) \div 4 =$
11. $81 \div 9 - 6 =$
12. $8 \times (11 + 7) =$
13. $(7 + 9) \times 9 =$
14. $65 - 26 - 24 =$
15. $96 \div (12 - 4) =$
16. $7 \times 9 + 12 =$
17. $16 + 84 \div 4 =$
18. $6 \times 8 \div 4 =$
19. $42 \div 3 \times 4 =$
20. $56 - 24 - 17 =$
21. $70 \div 5 \div 2 =$
22. $3 \times 5 \times 8 =$
23. $(20 + 16) \div 2 =$
24. $(73 - 7) \div 11 =$

1. $37 + 11 - 25 =$
2. $54 \div 2 + 27 =$
3. $57 - 8 \times 7 =$
4. $7 \times 9 - 17 =$
5. $8 \times (13 - 7) =$
6. $66 \div (6 + 5) =$
7. $35 + 15 + 25 =$
8. $(43 - 26) \times 6 =$
9. $4 \times (21 - 12) =$
10. $3 \times (24 + 35) =$
11. $48 \div 2 - 17 =$
12. $48 - 16 \div 4 =$
13. $13 + 7 \times 7 =$
14. $100 - 35 - 47 =$
15. $(5 + 9) \times 7 =$
16. $13 \times 9 \div 3 =$
17. $98 \div (36 - 22) =$
18. $7 \times 3 + 59 =$
19. $65 \div 13 \times 12 =$
20. $4 \times 5 \times 7 =$
21. $(46 + 62) \div 9 =$
22. $76 - 28 + 29 =$
23. $126 \div 3 \div 3 =$
24. $(60 - 12) \div 8 =$

1. $37 + 16 - 23 =$

2. $7 \times 7 - 18 =$

3. $43 - 8 \times 4 =$

4. $54 \div 6 - 3 =$

5. $11 + 22 + 33 =$

6. $(31 - 13) \times 5 =$

7. $6 \times (21 - 8) =$

8. $48 - 48 \div 8 =$

9. $7 + 7 \times 8 =$

10. $36 \div (9 + 9) =$

11. $69 \div 3 - 4 =$

12. $67 - 28 - 17 =$

13. $(31 + 42) \times 3 =$

14. $5 \times (21 + 28) =$

15. $125 \div 5 - 8 =$

16. $11 \times 9 + 8 =$

17. $45 \div (12 - 7) =$

18. $45 - 21 + 14 =$

19. $65 \div 5 \times 8 =$

20. $25 \times 12 \div 3 =$

21. $(24 + 36) \div 4 =$

22. $4 \times 6 \times 6 =$

23. $98 \div 7 \div 2 =$

24. $(63 - 9) \div 2 =$

1. $26 + 15 - 13 =$
2. $8 \times 4 - 19 =$
3. $50 - (4 \times 8) =$
4. $72 \div 2 + 27 =$
5. $31 + 17 + 24 =$
6. $(25 - 8) \times 7 =$
7. $7 \times (33 - 18) =$
8. $160 \div (8 + 2) =$
9. $70 \div (14 - 9) =$
10. $30 - (30 \div 6) =$
11. $25 + 3 \times 8 =$
12. $7 \times (12 + 14) =$
13. $96 \div (22 - 6) =$
14. $123 - 32 - 44 =$
15. $(12 + 13) \times 6 =$
16. $7 \times 9 + 36 =$
17. $56 \div 8 \times 9 =$
18. $11 \times 12 \div 12 =$
19. $9 + 78 \div 13 =$
20. $73 - 15 + 27 =$
21. $153 \div 3 \div 3 =$
22. $5 \times 6 \times 7 =$
23. $(37 + 73) \div 5 =$
24. $(55 - 19) \div 6 =$

1. $28 + 14 - 17 =$
2. $6 \times 7 - 27 =$
3. $80 - 7 \times 8 =$
4. $81 \div 9 + 18 =$
5. $24 + 35 + 46 =$
6. $(31 - 19) \times 5 =$
7. $6 \times (32 - 17) =$
8. $162 \div (6 + 3) =$
9. $45 + 8 \times 3 =$
10. $35 - 63 \div 7 =$
11. $120 \div (32 - 8) =$
12. $10 \times (17 + 19) =$
13. $60 \div 4 - 4 =$
14. $120 - 35 - 48 =$
15. $(9 + 8) \times 8 =$
16. $12 \times 7 + 21 =$
17. $64 \div (34 - 18) =$
18. $16 \times 8 \div 4 =$
19. $16 + 121 \div 11 =$
20. $66 - 32 - 34 =$
21. $96 \div 2 \div 4 =$
22. $9 \times 2 \times 6 =$
23. $(55 + 35) \div 18 =$
24. $(100 - 28) \div 8 =$

1. $29 + 14 - 17 =$

2. $5 \times 6 - 9 =$

3. $78 - 9 \times 5 =$

4. $99 \div 9 + 13 =$

5. $9 \times (11 - 2) =$

6. $(14 - 7) \times 13 =$

7. $53 + 26 + 55 =$

8. $54 \div (12 + 6) =$

9. $105 \div 5 - 13 =$

10. $22 - 35 \div 7 =$

11. $55 + 4 \times 4 =$

12. $67 - 32 - 19 =$

13. $64 \div (12 - 4) =$

14. $8 \times 12 + 35 =$

15. $(12 + 24) \times 8 =$

16. $128 \div 4 \div 2 =$

17. $135 \div 9 \div 5 =$

18. $55 - 18 + 22 =$

19. $150 \div 5 \times 4 =$

20. $9 \times 8 + 71 =$

21. $39 + 39 \div 3 =$

22. $7 \times 5 \times 9 =$

23. $(45 + 27) \div 3 =$

24. $(90 - 36) \div 3 =$

1. $21 + 19 - 6 =$

2. $5 \times 5 - 6 =$

3. $54 - 9 \times 3 =$

4. $51 \div 3 + 18 =$

5. $33 + 8 + 7 =$

6. $(32 - 18) \times 5 =$

7. $12 \times (7 - 6) =$

8. $45 \div (8 + 7) =$

9. $19 + 7 \times 8 =$

10. $81 - 7 \times 6 =$

11. $72 \div 6 - 5 =$

12. $6 \times (14 + 6) =$

13. $(12 + 13) \times 5 =$

14. $51 - 12 - 15 =$

15. $98 \div (11 - 4) =$

16. $6 \times 8 + 44 =$

17. $100 \div (33 - 8) =$

18. $48 \times 6 \div 2 =$

19. $15 + 49 \div 7 =$

20. $73 - 28 + 31 =$

21. $200 \div 5 \div 8 =$

22. $5 \times 5 \times 5 =$

23. $(55 + 17) \div 9 =$

24. $(54 - 9) \div 5 =$

1. $43 + 12 + 27 =$
2. $5 \times 6 - 11 =$
3. $44 - 5 \times 7 =$
4. $36 \div 6 + 26 =$
5. $24 + 27 + 38 =$
6. $(22 - 11) \times 11 =$
7. $9 \times (13 - 4) =$
8. $51 \div (7 - 4) =$
9. $35 + 7 \times 8 =$
10. $15 - 54 \div 6 =$
11. $(17 + 19) \times 7 =$
12. $6 \times (7 + 7) =$
13. $135 \div 9 + 6 =$
14. $92 - 24 - 49 =$
15. $111 \div (9 - 6) =$
16. $7 \times 7 + 7 =$
17. $45 + 36 \div 3 =$
18. $9 \times 8 \div 3 =$
19. $72 \div 6 \times 8 =$
20. $57 - 23 - 19 =$
21. $96 \div 3 \div 4 =$
22. $8 \times 7 \times 6 =$
23. $(22 + 33) \div 5 =$
24. $(86 - 8) \div 6 =$

Round the following numbers to the nearest ten.

Example: 547 rounded to the nearest ten is 550.

1. 128 = ________ 2. 244 = ________ 3. 791 = ________

4. 517 = ________ 5. 536 = ________ 6. 951 = ________

7. 449 = ________ 8. 213 = ________ 9. 221 = ________

10. 572 = ________ 11. 385 = ________ 12. 370 = ________

13. 388 = ________ 14. 739 = ________ 15. 108 = ________

16. 315 = ________ 17. 144 = ________ 18. 935 = ________

19. 257 = ________ 20. 348 = ________ 21. 238 = ________

22. 333 = ________ 23. 989 = ________ 24. 357 = ________

25. 421 = ________ 26. 201 = ________ 27. 549 = ________

Round the following numbers to the nearest ten.

Example: 547 rounded to the nearest ten is 550.

1. 855 = ________ 2. 618 = ________ 3. 696 = ________

4. 977 = ________ 5. 205 = ________ 6. 985 = ________

7. 279 = ________ 8. 984 = ________ 9. 632 = ________

10. 275 = ________ 11. 724 = ________ 12. 313 = ________

13. 122 = ________ 14. 302 = ________ 15. 117 = ________

16. 253 = ________ 17. 250 = ________ 18. 241 = ________

19. 308 = ________ 20. 946 = ________ 21. 853 = ________

22. 962 = ________ 23. 925 = ________ 24. 374 = ________

25. 837 = ________ 26. 114 = ________ 27. 606 = ________

Name: **Lesson 10-3** **Rounding numbers**

Round the following numbers to the nearest ten.

Example: 547 rounded to the nearest ten is 550.

1. 696 = ________	2. 638 = ________	3. 125 = ________
4. 701 = ________	5. 929 = ________	6. 450 = ________
7. 882 = ________	8. 674 = ________	9. 467 = ________
10. 377 = ________	11. 799 = ________	12. 286 = ________
13. 118 = ________	14. 552 = ________	15. 817 = ________
16. 451 = ________	17. 338 = ________	18. 981 = ________
19. 979 = ________	20. 219 = ________	21. 492 = ________
22. 698 = ________	23. 622 = ________	24. 404 = ________
25. 544 = ________	26. 881 = ________	27. 229 = ________

Round the following numbers to the nearest ten.
Example: 547 rounded to the nearest ten is 550.

1. $617 =$ ______	2. $867 =$ ______	3. $667 =$ ______
4. $807 =$ ______	5. $257 =$ ______	6. $598 =$ ______
7. $773 =$ ______	8. $717 =$ ______	9. $277 =$ ______
10. $464 =$ ______	11. $468 =$ ______	12. $920 =$ ______
13. $925 =$ ______	14. $477 =$ ______	15. $836 =$ ______
16. $624 =$ ______	17. $213 =$ ______	18. $256 =$ ______
19. $334 =$ ______	20. $362 =$ ______	21. $725 =$ ______
22. $455 =$ ______	23. $635 =$ ______	24. $247 =$ ______
25. $805 =$ ______	26. $138 =$ ______	27. $101 =$ ______

Round the following numbers to the nearest ten.

Example: 547 rounded to the nearest ten is 550.

1. 731 = ________ 2. 814 = ________ 3. 717 = ________

4. 222 = ________ 5. 729 = ________ 6. 942 = ________

7. 594 = ________ 8. 389 = ________ 9. 729 = ________

10. 819 = ________ 11. 367 = ________ 12. 132 = ________

13. 936 = ________ 14. 233 = ________ 15. 377 = ________

16. 492 = ________ 17. 187 = ________ 18. 567 = ________

19. 402 = ________ 20. 236 = ________ 21. 583 = ________

22. 645 = ________ 23. 784 = ________ 24. 338 = ________

25. 352 = ________ 26. 452 = ________ 27. 223 = ________

Round the following numbers to the nearest hundred.

Example: 678 rounded to the nearest hundred is 700.

1. 897 = ________ 2. 791 = ________ 3. 237 = ________

4. 191 = ________ 5. 369 = ________ 6. 487 = ________

7. 522 = ________ 8. 507 = ________ 9. 972 = ________

10. 327 = ________ 11. 487 = ________ 12. 318 = ________

13. 962 = ________ 14. 917 = ________ 15. 501 = ________

16. 724 = ________ 17. 399 = ________ 18. 405 = ________

19. 648 = ________ 20. 175 = ________ 21. 233 = ________

22. 697 = ________ 23. 432 = ________ 24. 176 = ________

25. 102 = ________ 26. 989 = ________ 27. 960 = ________

Round the following numbers to the nearest hundred.

Example: 678 rounded to the nearest hundred is 700.

1. 2227 = ________ 2. 735 = ________ 3. 5937 = ________

4. 1566 = ________ 5. 3512 = ________ 6. 408 = ________

7. 943 = ________ 8. 7292 = ________ 9. 155 = ________

10. 887 = ________ 11. 4037 = ________ 12. 3852 = ________

13. 3856 = ________ 14. 405 = ________ 15. 2840 = ________

16. 627 = ________ 17. 138 = ________ 18. 1107 = ________

19. 2247 = ________ 20. 1536 = ________ 21. 117 = ________

22. 5564 = ________ 23. 793 = ________ 24. 1220 = ________

25. 610 = ________ 26. 4270 = ________ 27. 538 = ________

Name: **Lesson 10-8** **Rounding numbers**

Round the following numbers to the nearest hundred.

Example: 678 rounded to the nearest hundred is 700.

1. 396 = ________	2. 8566 = ________	3. 108 = ________
4. 892 = ________	5. 711 = ________	6. 779 = ________
7. 4581 = ________	8. 160 = ________	9. 5158 = ________
10. 2401 = ________	11. 7452 = ________	12. 505 = ________
13. 365 = ________	14. 3572 = ________	15. 8234 = ________
16. 529 = ________	17. 226 = ________	18. 9477 = ________
19. 741 = ________	20. 8030 = ________	21. 7174 = ________
22. 8559 = ________	23. 149 = ________	24. 928 = ________
25. 2516 = ________	26. 9947 = ________	27. 607 = ________

Name: **Lesson 10-9** **Rounding numbers**

Round the following numbers to the nearest thousand.

Example: 1785 rounded to the nearest thousand is 2000.

1. 1431 = ________	2. 4115 = ________	3. 913 = ________
4. 5630 = ________	5. 843 = ________	6. 8938 = ________
7. 6557 = ________	8. 8959 = ________	9. 7918 = ________
10. 7749 = ________	11. 639 = ________	12. 6927 = ________
13. 7415 = ________	14. 3502 = ________	15. 1713 = ________
16. 9728 = ________	17. 2860 = ________	18. 4358 = ________
19. 8242 = ________	20. 922 = ________	21. 6324 = ________
22. 8903 = ________	23. 5187 = ________	24. 4276 = ________
25. 4834 = ________	26. 8124 = ________	27. 2518 = ________

Round the following numbers to the nearest thousand.

Example: 1785 rounded to the nearest thousand is 2000.

1. 3296 = ________	2. 35675 = ________	3. 5673 = ________
4. 6754 = ________	5. 7190 = ________	6. 2145 = ________
7. 19052 = ________	8. 86609 = ________	9. 32494 = ________
10. 42849 = ________	11. 7198 = ________	12. 6630 = ________
13. 9963 = ________	14. 2759 = ________	15. 89430 = ________
16. 16660 = ________	17. 41078 = ________	18. 6544 = ________
19. 2227 = ________	20. 77796 = ________	21. 38186 = ________
22. 6884 = ________	23. 2149 = ________	24. 4517 = ________
25. 14890 = ________	26. 3481 = ________	27. 2592 = ________

Name: **Lesson 11-1** **Finding the average**

Find the sum and average of the following numbers. For example,

The sum of 24 and 48 is $24 + 48 = 72$.
The average of 24 and 48 is $\frac{24+48}{2} = \frac{72}{2} = 36$.

1. 24, 48 • Sum = <u>72</u> • Average = <u>36</u>
2. 17, 33 • Sum = _____ • Average = _____
3. 42, 26 • Sum = _____ • Average = _____
4. 36, 42 • Sum = _____ • Average = _____
5. 11, 99 • Sum = _____ • Average = _____
6. 33, 53 • Sum = _____ • Average = _____
7. 17, 59 • Sum = _____ • Average = _____
8. 55, 75 • Sum = _____ • Average = _____
9. 86, 68 • Sum = _____ • Average = _____
10. 22, 82 • Sum = _____ • Average = _____
11. 51, 57 • Sum = _____ • Average = _____

Name: **Lesson 11-2 Finding the average**

Find the sum and average of the following numbers. For example,

The sum of 24 and 48 is $24 + 48 = 72$.
The average of 24 and 48 is $\frac{24+48}{2} = \frac{72}{2} = 36$.

1. 16, 88 • Sum = _____ • Average = _____
2. 86, 58 • Sum = _____ • Average = _____
3. 52, 74 • Sum = _____ • Average = _____
4. 35, 65 • Sum = _____ • Average = _____
5. 53, 11 • Sum = _____ • Average = _____
6. 98, 56 • Sum = _____ • Average = _____
7. 38, 90 • Sum = _____ • Average = _____
8. 27, 27 • Sum = _____ • Average = _____
9. 43, 91 • Sum = _____ • Average = _____
10. 64, 46 • Sum = _____ • Average = _____
11. 15, 93 • Sum = _____ • Average = _____

Name: **Lesson 11-3 Finding the average**

Find the sum and average of the following numbers. For example,

The sum of 24 and 48 is $24 + 48 = 72$.
The average of 24 and 48 is $\frac{24+48}{2} = \frac{72}{2} = 36$.

1. 83, 51 • Sum = _____ • Average = _____

2. 10, 86 • Sum = _____ • Average = _____

3. 79, 39 • Sum = _____ • Average = _____

4. 84, 36 • Sum = _____ • Average = _____

5. 85, 69 • Sum = _____ • Average = _____

6. 91, 67 • Sum = _____ • Average = _____

7. 16, 90 • Sum = _____ • Average = _____

8. 46, 10 • Sum = _____ • Average = _____

9. 44, 12 • Sum = _____ • Average = _____

10. 42, 68 • Sum = _____ • Average = _____

11. 62, 38 • Sum = _____ • Average = _____

Name: **Lesson 11-4 Finding the average**

Find the sum and average of the following numbers. For example,

The sum of 24 and 48 is $24 + 48 = 72$.
The average of 24 and 48 is $\frac{24+48}{2} = \frac{72}{2} = 36$.

1. 33, 39 • Sum = _____ • Average = _____

2. 25, 59 • Sum = _____ • Average = _____

3. 55, 27 • Sum = _____ • Average = _____

4. 69, 41 • Sum = _____ • Average = _____

5. 49, 57 • Sum = _____ • Average = _____

6. 27, 47 • Sum = _____ • Average = _____

7. 37, 87 • Sum = _____ • Average = _____

8. 89, 27 • Sum = _____ • Average = _____

9. 46, 68 • Sum = _____ • Average = _____

10. 28, 66 • Sum = _____ • Average = _____

11. 88, 98 • Sum = _____ • Average = _____

Name: **Lesson 11-5 Finding the average**

Find the sum and average of the following numbers. For example,
The sum of 24 and 48 is $24 + 48 = 72$.
The average of 24 and 48 is $\frac{24+48}{2} = \frac{72}{2} = 36$.

1. 38, 36 • Sum = _____ • Average = _____
2. 86, 12 • Sum = _____ • Average = _____
3. 61, 91 • Sum = _____ • Average = _____
4. 25, 43 • Sum = _____ • Average = _____
5. 44, 76 • Sum = _____ • Average = _____
6. 10, 24 • Sum = _____ • Average = _____
7. 37, 85 • Sum = _____ • Average = _____
8. 47, 81 • Sum = _____ • Average = _____
9. 80, 32 • Sum = _____ • Average = _____
10. 40, 28 • Sum = _____ • Average = _____
11. 30, 66 • Sum = _____ • Average = _____

Name: **Lesson 11-6 Finding the average**

Find the sum and average of the following numbers. For example,

The sum of 24 and 48 is $24 + 48 = 72$.
The average of 24 and 48 is $\frac{24+48}{2} = \frac{72}{2} = 36$.

1. 89, 91 • Sum = _____ • Average = _____

2. 90, 28 • Sum = _____ • Average = _____

3. 113, 43 • Sum = _____ • Average = _____

4. 69, 57 • Sum = _____ • Average = _____

5. 103, 87 • Sum = _____ • Average = _____

6. 88, 96 • Sum = _____ • Average = _____

7. 52, 56 • Sum = _____ • Average = _____

8. 74, 128 • Sum = _____ • Average = _____

9. 114, 76 • Sum = _____ • Average = _____

10. 94, 118 • Sum = _____ • Average = _____

11. 14, 138 • Sum = _____ • Average = _____

Name: **Lesson 11-7 Finding the average**

Find the sum and average of the following numbers. For example,
The sum of 24 and 48 is $24 + 48 = 72$.
The average of 24 and 48 is $\frac{24+48}{2} = \frac{72}{2} = 36$.

1. 130, 112 • Sum = _____ • Average = _____
2. 127, 105 • Sum = _____ • Average = _____
3. 117, 139 • Sum = _____ • Average = _____
4. 39, 131 • Sum = _____ • Average = _____
5. 60, 28 • Sum = _____ • Average = _____
6. 19, 59 • Sum = _____ • Average = _____
7. 129, 141 • Sum = _____ • Average = _____
8. 47, 107 • Sum = _____ • Average = _____
9. 54, 146 • Sum = _____ • Average = _____
10. 45, 61 • Sum = _____ • Average = _____
11. 119, 151 • Sum = _____ • Average = _____

Name: **Lesson 11-8 Finding the average**

Find the sum and average of the following numbers. For example,

The sum of 24 and 48 is $24 + 48 = 72$.
The average of 24 and 48 is $\frac{24+48}{2} = \frac{72}{2} = 36$.

1. 69, 153 • Sum = ____ • Average = ____
2. 86, 104 • Sum = ____ • Average = ____
3. 22, 154 • Sum = ____ • Average = ____
4. 159, 69 • Sum = ____ • Average = ____
5. 61, 55 • Sum = ____ • Average = ____
6. 145, 55 • Sum = ____ • Average = ____
7. 37, 79 • Sum = ____ • Average = ____
8. 63, 151 • Sum = ____ • Average = ____
9. 36, 42 • Sum = ____ • Average = ____
10. 31, 117 • Sum = ____ • Average = ____
11. 99, 81 • Sum = ____ • Average = ____

Name: **Lesson 11-9** **Finding the average**

Find the sum and average of the following numbers. For example,
The sum of 24 and 48 is $24 + 48 = 72$.
The average of 24 and 48 is $\frac{24+48}{2} = \frac{72}{2} = 36$.

1. 177, 45 • Sum = _____ • Average = _____
2. 103, 45 • Sum = _____ • Average = _____
3. 90, 118 • Sum = _____ • Average = _____
4. 125, 101 • Sum = _____ • Average = _____
5. 37, 17 • Sum = _____ • Average = _____
6. 94, 144 • Sum = _____ • Average = _____
7. 43, 101 • Sum = _____ • Average = _____
8. 164, 132 • Sum = _____ • Average = _____
9. 102, 130 • Sum = _____ • Average = _____
10. 35, 109 • Sum = _____ • Average = _____
11. 31, 119 • Sum = _____ • Average = _____

Name: **Lesson 11-10** **Finding the average**

Find the sum and average of the following numbers. For example,
The sum of 24 and 48 is $24 + 48 = 72$.
The average of 24 and 48 is $\frac{24+48}{2} = \frac{72}{2} = 36$.

1. 158, 68 • Sum = _____ • Average = _____
2. 176, 14 • Sum = _____ • Average = _____
3. 129, 151 • Sum = _____ • Average = _____
4. 92, 146 • Sum = _____ • Average = _____
5. 162, 188 • Sum = _____ • Average = _____
6. 57, 75 • Sum = _____ • Average = _____
7. 171, 123 • Sum = _____ • Average = _____
8. 118, 168 • Sum = _____ • Average = _____
9. 71, 57 • Sum = _____ • Average = _____
10. 34, 144 • Sum = _____ • Average = _____
11. 136, 58 • Sum = _____ • Average = _____

Name: **Lesson 12-1 Counting the total value of coins**

Count the total value of the following coins in cents.

Example: The total value of 1 quarters, 1 dime, 1 nickel, and 1 penny is $25 + 10 + 5 + 1 = 41$ cents.

1. 2 quarters, 1 dime, and 1 penny = ________ cents

2. 1 quarter, 3 dimes, and 2 pennies = ________ cents

3. 2 quarters, 1 nickel, and 3 pennies = ________ cents

4. 1 quarter, 2 dimes, and 1 nickel = ________ cents

5. 2 quarters, 1 dime, and 2 nickels = ________ cents

6. 1 quarter, 3 dimes, and 1 nickel = ________ cents

7. 3 quarters, 1 dimes, and 1 nickel = ________ cents

8. 3 quarters, 1 dime, and 3 pennies = ________ cents

9. 3 quarters, 2 nickels, and 2 pennies = ________ cents

10. 3 quarters, 3 dimes, and 2 pennies = ________ cents

11. 3 quarters, 3 dimes, and 3 nickels = ________ cents

Name: **Lesson 12-2 Counting the total value of coins**

Count the total value of the following coins in cents.

Example: The total value of 1 quarters, 1 dime, 1 nickel, and 1 penny is $25 + 10 + 5 + 1 = 41$ cents.

1. 2 quarters, 2 dimes, and 2 pennies = ________ cents
2. 2 quarters, 4 dimes, and 3 nickels = ________ cents
3. 3 quarters, 1 dime, and 2 nickels = ________ cents
4. 2 quarters, 4 nickels, and 3 pennies = ________ cents
5. 3 quarters, 2 dimes, and 1 nickel = ________ cents
6. 4 quarters, 2 dimes, and 3 nickels = ________ cents
7. 3 quarters, 4 dimes, and 1 penny = ________ cents
8. 3 quarters, 3 nickels, and 3 pennies = ________ cents
9. 4 quarters, 3 dimes, and 4 nickels = ________ cents
10. 4 quarters, 4 dimes, and 3 nickels = ________ cents
11. 2 quarters, 3 nickels, and 1 penny = ________ cents

Name: **Lesson 12-3 Counting the total value of coins**

Count the total value of the following coins in cents.

Example: The total value of 1 quarters, 1 dime, 1 nickel, and 1 penny is $25 + 10 + 5 + 1 = 41$ cents.

1. 3 quarters, 2 nickels, and 2 pennies = __________ cents

2. 2 quarters, 4 dimes, and 3 nickels = __________ cents

3. 4 quarters, 1 nickel, and 3 pennies = __________ cents

4. 4 quarters, 2 nickels, and 1 penny = __________ cents

5. 4 quarters, 4 nickels, and 3 pennies = __________ cents

6. 3 quarters, 4 nickels, and 3 pennies = __________ cents

7. 2 quarters, 3 dimes, and 3 nickels = __________ cents

8. 4 quarters, 3 nickels, and 3 pennies = __________ cents

9. 3 quarters, 3 nickels, and 4 pennies = __________ cents

10. 4 quarters, 3 dimes, and 2 pennies = __________ cents

11. 2 quarters, 3 dimes, and 4 nickels = __________ cents

Name: **Lesson 12-4 Counting the total value of coins**

Count the total value of the following coins in cents.

Example: The total value of 1 quarters, 1 dime, 1 nickel, and 1 penny is $25 + 10 + 5 + 1 = 41$ cents.

1. 4 quarters, 4 dimes, and 4 pennies = __________ cents

2. 4 quarters, 1 dime, and 1 nickel = __________ cents

3. 3 quarters, 3 dimes, and 1 penny = __________ cents

4. 2 quarters, 4 dimes, and 3 pennies = __________ cents

5. 4 quarters, 3 dimes, and 3 pennies = __________ cents

6. 4 quarters, 2 dimes, and 2 nickels = __________ cents

7. 3 quarters, 4 dimes, and 2 pennies = __________ cents

8. 4 quarters, 1 dime, and 4 pennies = __________ cents

9. 3 quarters, 1 dime, and 2 pennies = __________ cents

10. 4 quarters, 3 dimes, and 2 nickels = __________ cents

11. 4 quarters, 2 dimes, and 1 penny = __________ cents

Name: **Lesson 12-5 Counting the total value of coins**

Count the total value of the following coins in cents.

Example: The total value of 1 quarters, 1 dime, 1 nickel, and 1 penny is $25 + 10 + 5 + 1 = 41$ cents.

1. 3 quarters, 4 dimes, and 3 pennies = ________ cents

2. 3 quarters, 3 dimes, and 5 nickels = ________ cents

3. 4 quarters, 3 dimes, and 3 pennies = ________ cents

4. 5 quarters, 3 nickels, and 5 pennies = ________ cents

5. 3 quarters, 5 dimes, and 4 pennies = ________ cents

6. 4 quarters, 5 dimes, and 3 pennies = ________ cents

7. 3 quarters, 5 dimes, and 3 nickels = ________ cents

8. 5 quarters, 4 nickels, and 3 pennies = ________ cents

9. 3 quarters, 2 dimes, and 1 penny = ________ cents

10. 4 quarters, 5 dimes, and 2 pennies = ________ cents

11. 3 quarters, 4 dimes, and 1 nickel = ________ cents

Name: **Lesson 12-6 Counting the total value of coins**

Count the total value of the following coins in cents.

Example: The total value of 1 quarters, 1 dime, 1 nickel, and 1 penny is $25 + 10 + 5 + 1 = 41$ cents.

1. 5 quarters, 1 nickel, and 2 pennies = ________ cents
2. 4 quarters, 5 dimes, and 1 nickel = ________ cents
3. 3 quarters, 4 dimes, and 4 pennies = ________ cents
4. 2 quarters, 3 dimes, and 5 pennies = ________ cents
5. 5 quarters, 4 nickels, and 3 pennies = ________ cents
6. 4 quarters, 4 dimes, and 2 nickels = ________ cents
7. 3 quarters, 5 dimes, and 4 pennies = ________ cents
8. 5 quarters, 5 nickels, and 4 pennies = ________ cents
9. 4 quarters, 3 dimes, and 5 nickels = ________ cents
10. 3 quarters, 1 dime, and 5 pennies = ________ cents
11. 5 quarters, 3 nickels, and 1 penny = ________ cents

Name: **Lesson 12-7 Counting the total value of coins**

Count the total value of the following coins in cents.

Example: The total value of 1 quarters, 1 dime, 1 nickel, and 1 penny is $25 + 10 + 5 + 1 = 41$ cents.

1. 5 quarters, 5 dimes, and 5 nickels = ________ cents

2. 6 quarters, 1 dime, and 3 pennies = ________ cents

3. 4 quarters, 4 nickels, and 1 penny = ________ cents

4. 5 quarters, 3 dimes, and 3 nickels = ________ cents

5. 6 quarters, 2 dimes, and 5 pennies = ________ cents

6. 5 quarters, 1 nickel, and 2 pennies = ________ cents

7. 5 quarters, 1 dime, and 3 nickels = ________ cents

8. 6 quarters, 3 dimes, and 3 pennies = ________ cents

9. 4 quarters, 2 nickels, and 6 pennies = ________ cents

10. 6 quarters, 6 dimes, and 6 pennies = ________ cents

11. 5 quarters, 4 dimes, and 2 nickels = ________ cents

Name: **Lesson 12-8 Counting the total value of coins**

Count the total value of the following coins in cents.

Example: The total value of 1 quarters, 1 dime, 1 nickel, and 1 penny is $25 + 10 + 5 + 1 = 41$ cents.

1. 5 quarters, 4 nickels, and 4 pennies = ________ cents

2. 6 quarters, 6 dimes, and 5 pennies = ________ cents

3. 4 quarters, 1 nickel, and 2 pennies = ________ cents

4. 3 quarters, 3 dimes, and 5 nickels = ________ cents

5. 5 quarters, 5 nickels, and 6 pennies = ________ cents

6. 6 quarters, 5 dimes, and 6 pennies = ________ cents

7. 4 quarters, 2 nickels, and 4 pennies = ________ cents

8. 3 quarters, 6 nickels, and 5 pennies = ________ cents

9. 6 quarters, 4 dimes, and 1 penny = ________ cents

10. 2 quarters, 5 nickels, and 2 pennies = ________ cents

11. 6 quarters, 2 dimes, and 3 pennies = ________ cents

Name: **Lesson 12-9 Counting the total value of coins**

Count the total value of the following coins in cents.

Example: The total value of 1 quarters, 1 dime, 1 nickel, and 1 penny is $25 + 10 + 5 + 1 = 41$ cents.

1. 7 quarters, 1 dime, and 4 pennies = ________ cents

2. 6 quarters, 7 nickels, and 1 penny = ________ cents

3. 5 quarters, 1 dime, and 7 nickels = ________ cents

4. 7 quarters, 3 dimes, and 7 pennies = ________ cents

5. 6 quarters, 6 nickels, and 7 pennies = ________ cents

6. 3 quarters, 2 dimes, and 5 nickels = ________ cents

7. 5 quarters, 3 dimes, and 6 nickels = ________ cents

8. 4 quarters, 4 dimes, and 6 pennies = ________ cents

9. 6 quarters, 5 nickels, and 5 pennies = ________ cents

10. 1 quarter, 7 dimes, and 3 nickels = ________ cents

11. 7 quarters, 6 dimes, and 3 pennies = ________ cents

Name: **Lesson 12-10 Counting the total value of coins**

Count the total value of the following coins in cents.

Example: The total value of 1 quarters, 1 dime, 1 nickel, and 1 penny is $25 + 10 + 5 + 1 = 41$ cents.

1. 7 quarters, 7 dimes, and 1 penny = ________ cents

2. 6 quarters, 1 dime, and 2 nickels = ________ cents

3. 5 quarters, 7 nickels, and 7 pennies = ________ cents

4. 4 quarters, 6 dimes, and 7 nickels = ________ cents

5. 3 quarters, 6 dimes, and 3 pennies = ________ cents

6. 6 quarters, 2 dimes, and 4 nickels = ________ cents

7. 5 quarters, 6 nickels, and 5 pennies = ________ cents

8. 2 quarters, 3 dimes, and 4 pennies = ________ cents

9. 1 quarter, 7 dimes, and 3 nickels = ________ cents

10. 5 quarters, 5 nickels, and 5 pennies = ________ cents

11. 7 quarters, 2 dimes, and 7 pennies = ________ cents

Name: Lesson 1-1 Multiplying two digit numbers

1.	87	× 22	174	174	1914
2.	34	× 63	102	204	2142
3.	39	× 42	78	156	1638
4.	69	× 52	138	345	3588
5.	39	× 63	117	234	2457
6.	67	× 32	134	201	2144
7.	97	× 79	873	679	7663
8.	48	× 83	144	384	3984
9.	77	× 28	616	154	2156
10.	48	× 72	96	336	3456
11.	35	× 44	140	140	1540
12.	98	× 55	490	490	5390
13.	85	× 64	340	510	5440
14.	47	× 34	188	141	1598
15.	74	× 84	296	592	6216
16.	99	× 98	792	891	9702

Name: Lesson 1-2 Multiplying two digit numbers

1.	98	× 22	196	196	2156
2.	79	× 32	158	237	2528
3.	61	× 42	122	244	2562
4.	89	× 52	178	445	4628
5.	79	× 62	158	474	4898
6.	68	× 72	136	476	4896
7.	99	× 86	594	792	8514
8.	57	× 82	114	456	4674
9.	88	× 28	704	176	2464
10.	62	× 34	248	186	2108
11.	85	× 47	595	340	3995
12.	48	× 56	288	240	2688
13.	47	× 65	235	282	3055
14.	53	× 75	265	371	3975
15.	49	× 77	343	343	3773
16.	96	× 96	576	864	9216

Name: Lesson 1-3 Multiplying two digit numbers

1.	92	× 23	276	184	2116
2.	98	× 32	196	294	3136
3.	46	× 53	138	230	2438
4.	72	× 37	504	216	2664
5.	94	× 62	188	564	5828
6.	85	× 81	85	680	6885
7.	49	× 95	245	441	4655
8.	44	× 33	132	132	1452
9.	67	× 29	603	134	1943
10.	79	× 43	237	316	3397
11.	58	× 67	406	348	3886
12.	39	× 97	273	351	3783
13.	58	× 66	348	348	3828
14.	85	× 75	425	595	6375
15.	94	× 99	846	846	9306
16.	73	× 76	438	511	5548

Name: Lesson 1-4 Multiplying two digit numbers

1.	93	× 23	279	186	2139
2.	35	× 45	175	140	1575
3.	42	× 76	252	294	3192
4.	89	× 53	267	445	4717
5.	45	× 73	135	315	3285
6.	96	× 72	192	672	6912
7.	98	× 89	882	784	8722
8.	43	× 92	86	387	3956
9.	77	× 39	693	231	3003
10.	81	× 34	324	243	2754
11.	64	× 65	320	384	4160
12.	74	× 59	666	370	4366
13.	64	× 74	256	448	4736
14.	94	× 73	282	658	6862
15.	76	× 88	608	608	6688
16.	95	× 87	665	760	8265

Name: **Lesson 1-5 Multiplying two digit numbers**

1.	97	× 33	291	291	3201
2.	37	× 42	74	148	1554
3.	99	× 42	198	396	4158
4.	91	× 55	455	455	5005
5.	59	× 63	177	354	3717
6.	82	× 72	164	574	5904
7.	98	× 83	294	784	8134
8.	58	× 99	522	522	5742
9.	89	× 37	623	267	3293
10.	95	× 43	285	380	4085
11.	36	× 63	108	216	2268
12.	84	× 65	420	504	5460
13.	78	× 65	390	468	5070
14.	86	× 97	602	774	8342
15.	74	× 84	296	592	6216
16.	89	× 89	801	712	7921

Name: **Lesson 1-6 Multiplying two digit numbers**

1.	96	× 39	864	288	3744
2.	74	× 33	222	222	2442
3.	52	× 43	156	208	2236
4.	49	× 54	196	245	2646
5.	78	× 63	234	468	4914
6.	52	× 73	156	364	3796
7.	97	× 83	291	776	8051
8.	64	× 98	512	576	6272
9.	79	× 38	632	237	3002
10.	59	× 36	354	177	2124
11.	57	× 46	342	228	2622
12.	57	× 57	399	285	3249
13.	93	× 65	465	558	6045
14.	95	× 78	760	665	7410
15.	88	× 76	528	616	6688
16.	89	× 96	534	801	8544

Name: **Lesson 1-7 Multiplying two digit numbers**

1.	98	× 34	392	294	3332
2.	78	× 36	468	234	2808
3.	72	× 43	216	288	3096
4.	75	× 53	225	375	3975
5.	94	× 63	282	564	5922
6.	59	× 73	177	413	4307
7.	87	× 81	87	696	7047
8.	68	× 99	612	612	6732
9.	99	× 32	198	297	3168
10.	39	× 36	234	117	1404
11.	77	× 44	308	308	3388
12.	56	× 59	504	280	3304
13.	95	× 66	570	570	6270
14.	87	× 89	783	696	7743
15.	86	× 85	430	688	7310
16.	76	× 91	76	684	6916

Name: **Lesson 1-8 Multiplying two digit numbers**

1.	85	× 42	170	340	3570
2.	49	× 57	343	245	2793
3.	73	× 25	365	146	1825
4.	79	× 36	474	237	2844
5.	97	× 27	679	194	2619
6.	85	× 72	170	595	6120
7.	64	× 85	320	512	5440
8.	54	× 49	486	216	2646
9.	83	× 56	498	415	4648
10.	48	× 63	144	288	3024
11.	78	× 65	390	468	5070
12.	96	× 85	480	768	8160
13.	97	× 77	679	679	7469
14.	56	× 77	392	392	4312
15.	89	× 98	712	801	8722
16.	86	× 39	774	258	3354

Name: **Lesson 1-9 Multiplying two digit numbers**

1. 99 × 33 297 297 3267	2. 69 × 24 276 138 1656	3. 87 × 43 261 348 3741	4. 94 × 54 376 470 5076
5. 43 × 64 172 258 2752	6. 92 × 38 736 276 3496	7. 72 × 39 648 216 2808	8. 65 × 95 325 585 6175
9. 89 × 37 623 267 3293	10. 84 × 47 588 336 3948	11. 85 × 84 340 680 7140	12. 85 × 69 765 510 5865
13. 87 × 85 435 696 7395	14. 97 × 75 485 679 7275	15. 91 × 99 819 819 9009	16. 87 × 98 696 783 8526

Name: **Lesson 1-10 Multiplying two digit numbers**

1. 76 × 27 532 152 2052	2. 53 × 59 477 265 3127	3. 76 × 86 456 608 6536	4. 57 × 51 57 285 2907
5. 65 × 63 195 390 4095	6. 85 × 55 425 425 4675	7. 71 × 71 71 497 5041	8. 74 × 93 222 666 6882
9. 79 × 81 79 632 6399	10. 87 × 74 348 609 6438	11. 86 × 94 344 774 8084	12. 56 × 88 448 448 4928
13. 75 × 79 675 525 5925	14. 62 × 97 434 558 6014	15. 94 × 78 752 658 7332	16. 97 × 66 582 582 6402

Name: **Lesson 2-1 Dividing by one digit numbers with remainders**

1. 21 2)43 40 3 2 1	2. 31 2)63 60 3 2 1	3. 19 3)59 30 29 27 2	4. 25 3)76 60 16 15 1
5. 9 4)39 36 3	6. 14 5)72 50 22 20 2	7. 17 5)88 50 38 35 3	8. 23 4)95 80 15 12 3
9. 16 3)50 30 20 18 2	10. 17 2)35 20 15 14 1	11. 16 3)49 30 19 18 1	12. 30 3)92 90 2
13. 23 3)70 60 10 9 1	14. 19 5)99 50 49 45 4	15. 13 6)83 60 23 18 5	16. 13 5)69 50 19 15 4
17. 12 4)51 40 11 8 3	18. 24 4)98 80 18 16 2	19. 10 6)63 60 3	20. 15 4)61 40 21 20 1

Name: **Lesson 2-2 Dividing by one digit numbers with remainders**

1. 18 2)37 20 17 16 1	2. 14 5)74 50 24 20 4	3. 17 3)52 30 22 21 1	4. 15 4)61 40 21 20 1
5. 24 4)97 80 17 16 1	6. 37 2)75 60 15 14 1	7. 19 5)99 50 49 45 4	8. 12 8)99 80 19 16 3
9. 21 2)43 40 3 2 1	10. 19 4)78 40 38 36 2	11. 23 3)70 60 10 9 1	12. 26 3)79 60 19 18 1
13. 12 3)37 30 7 6 1	14. 43 2)87 80 7 6 1	15. 12 6)77 60 17 12 5	16. 17 4)69 40 29 28 1
17. 13 5)66 50 16 15 1	18. 16 5)82 50 32 30 2	19. 28 3)85 60 25 24 1	20. 37 2)75 60 15 14 1

Name: **Lesson 2-3 Dividing by one digit numbers with remainders**

1. $\begin{array}{r} 15 \\ 2\overline{)31} \\ 20 \\ \hline 11 \\ 10 \\ \hline 1 \end{array}$

2. $\begin{array}{r} 15 \\ 5\overline{)78} \\ 50 \\ \hline 28 \\ 25 \\ \hline 3 \end{array}$

3. $\begin{array}{r} 20 \\ 3\overline{)61} \\ 60 \\ \hline 1 \end{array}$

4. $\begin{array}{r} 16 \\ 6\overline{)99} \\ 60 \\ \hline 39 \\ 36 \\ \hline 3 \end{array}$

5. $\begin{array}{r} 22 \\ 4\overline{)91} \\ 80 \\ \hline 11 \\ 8 \\ \hline 3 \end{array}$

6. $\begin{array}{r} 29 \\ 3\overline{)89} \\ 60 \\ \hline 29 \\ 27 \\ \hline 2 \end{array}$

7. $\begin{array}{r} 14 \\ 6\overline{)87} \\ 60 \\ \hline 27 \\ 24 \\ \hline 3 \end{array}$

8. $\begin{array}{r} 31 \\ 3\overline{)95} \\ 90 \\ \hline 5 \\ 3 \\ \hline 2 \end{array}$

9. $\begin{array}{r} 19 \\ 2\overline{)39} \\ 20 \\ \hline 19 \\ 18 \\ \hline 1 \end{array}$

10. $\begin{array}{r} 20 \\ 4\overline{)83} \\ 80 \\ \hline 3 \end{array}$

11. $\begin{array}{r} 29 \\ 3\overline{)89} \\ 60 \\ \hline 29 \\ 27 \\ \hline 2 \end{array}$

12. $\begin{array}{r} 10 \\ 7\overline{)76} \\ 70 \\ \hline 6 \end{array}$

13. $\begin{array}{r} 11 \\ 5\overline{)57} \\ 50 \\ \hline 7 \\ 5 \\ \hline 2 \end{array}$

14. $\begin{array}{r} 17 \\ 3\overline{)53} \\ 30 \\ \hline 23 \\ 21 \\ \hline 2 \end{array}$

15. $\begin{array}{r} 14 \\ 6\overline{)88} \\ 60 \\ \hline 28 \\ 24 \\ \hline 4 \end{array}$

16. $\begin{array}{r} 21 \\ 3\overline{)64} \\ 60 \\ \hline 4 \\ 3 \\ \hline 1 \end{array}$

17. $\begin{array}{r} 39 \\ 2\overline{)79} \\ 60 \\ \hline 19 \\ 18 \\ \hline 1 \end{array}$

18. $\begin{array}{r} 10 \\ 6\overline{)64} \\ 60 \\ \hline 4 \end{array}$

19. $\begin{array}{r} 25 \\ 3\overline{)76} \\ 60 \\ \hline 16 \\ 15 \\ \hline 1 \end{array}$

20. $\begin{array}{r} 20 \\ 2\overline{)41} \\ 40 \\ \hline 1 \end{array}$

Name: **Lesson 2-4 Dividing by one digit numbers with remainders**

1. $\begin{array}{r} 29 \\ 2\overline{)59} \\ 40 \\ \hline 19 \\ 18 \\ \hline 1 \end{array}$

2. $\begin{array}{r} 42 \\ 2\overline{)85} \\ 80 \\ \hline 5 \\ 4 \\ \hline 1 \end{array}$

3. $\begin{array}{r} 11 \\ 6\overline{)67} \\ 60 \\ \hline 7 \\ 6 \\ \hline 1 \end{array}$

4. $\begin{array}{r} 14 \\ 7\overline{)99} \\ 70 \\ \hline 29 \\ 28 \\ \hline 1 \end{array}$

5. $\begin{array}{r} 21 \\ 4\overline{)87} \\ 80 \\ \hline 7 \\ 4 \\ \hline 3 \end{array}$

6. $\begin{array}{r} 17 \\ 5\overline{)87} \\ 50 \\ \hline 37 \\ 35 \\ \hline 2 \end{array}$

7. $\begin{array}{r} 20 \\ 3\overline{)61} \\ 60 \\ \hline 1 \end{array}$

8. $\begin{array}{r} 24 \\ 4\overline{)97} \\ 80 \\ \hline 17 \\ 16 \\ \hline 1 \end{array}$

9. $\begin{array}{r} 36 \\ 2\overline{)73} \\ 60 \\ \hline 13 \\ 12 \\ \hline 1 \end{array}$

10. $\begin{array}{r} 49 \\ 2\overline{)99} \\ 80 \\ \hline 19 \\ 18 \\ \hline 1 \end{array}$

11. $\begin{array}{r} 15 \\ 5\overline{)78} \\ 50 \\ \hline 28 \\ 25 \\ \hline 3 \end{array}$

12. $\begin{array}{r} 11 \\ 7\overline{)78} \\ 70 \\ \hline 8 \\ 7 \\ \hline 1 \end{array}$

13. $\begin{array}{r} 14 \\ 5\overline{)71} \\ 50 \\ \hline 21 \\ 20 \\ \hline 1 \end{array}$

14. $\begin{array}{r} 15 \\ 6\overline{)95} \\ 60 \\ \hline 35 \\ 30 \\ \hline 5 \end{array}$

15. $\begin{array}{r} 18 \\ 3\overline{)55} \\ 30 \\ \hline 25 \\ 24 \\ \hline 1 \end{array}$

16. $\begin{array}{r} 32 \\ 3\overline{)98} \\ 90 \\ \hline 8 \\ 6 \\ \hline 2 \end{array}$

17. $\begin{array}{r} 19 \\ 3\overline{)58} \\ 30 \\ \hline 28 \\ 27 \\ \hline 1 \end{array}$

18. $\begin{array}{r} 40 \\ 2\overline{)81} \\ 80 \\ \hline 1 \end{array}$

19. $\begin{array}{r} 13 \\ 5\overline{)69} \\ 50 \\ \hline 19 \\ 15 \\ \hline 4 \end{array}$

20. $\begin{array}{r} 16 \\ 4\overline{)65} \\ 40 \\ \hline 25 \\ 24 \\ \hline 1 \end{array}$

Name: **Lesson 2-5 Dividing by one digit numbers with remainders**

1. $\begin{array}{r} 15 \\ 3\overline{)47} \\ 30 \\ \hline 17 \\ 15 \\ \hline 2 \end{array}$

2. $\begin{array}{r} 14 \\ 6\overline{)86} \\ 60 \\ \hline 26 \\ 24 \\ \hline 2 \end{array}$

3. $\begin{array}{r} 25 \\ 3\overline{)77} \\ 60 \\ \hline 17 \\ 15 \\ \hline 2 \end{array}$

4. $\begin{array}{r} 13 \\ 7\overline{)96} \\ 70 \\ \hline 26 \\ 21 \\ \hline 5 \end{array}$

5. $\begin{array}{r} 12 \\ 5\overline{)62} \\ 50 \\ \hline 12 \\ 10 \\ \hline 2 \end{array}$

6. $\begin{array}{r} 12 \\ 3\overline{)38} \\ 30 \\ \hline 8 \\ 6 \\ \hline 2 \end{array}$

7. $\begin{array}{r} 14 \\ 6\overline{)88} \\ 60 \\ \hline 28 \\ 24 \\ \hline 4 \end{array}$

8. $\begin{array}{r} 19 \\ 4\overline{)77} \\ 40 \\ \hline 37 \\ 36 \\ \hline 1 \end{array}$

9. $\begin{array}{r} 19 \\ 3\overline{)58} \\ 30 \\ \hline 28 \\ 27 \\ \hline 1 \end{array}$

10. $\begin{array}{r} 18 \\ 5\overline{)93} \\ 50 \\ \hline 43 \\ 40 \\ \hline 3 \end{array}$

11. $\begin{array}{r} 11 \\ 4\overline{)46} \\ 40 \\ \hline 6 \\ 4 \\ \hline 2 \end{array}$

12. $\begin{array}{r} 28 \\ 2\overline{)57} \\ 40 \\ \hline 17 \\ 16 \\ \hline 1 \end{array}$

13. $\begin{array}{r} 12 \\ 5\overline{)61} \\ 50 \\ \hline 11 \\ 10 \\ \hline 1 \end{array}$

14. $\begin{array}{r} 16 \\ 4\overline{)65} \\ 40 \\ \hline 25 \\ 24 \\ \hline 1 \end{array}$

15. $\begin{array}{r} 12 \\ 7\overline{)87} \\ 70 \\ \hline 17 \\ 14 \\ \hline 3 \end{array}$

16. $\begin{array}{r} 11 \\ 7\overline{)81} \\ 70 \\ \hline 11 \\ 7 \\ \hline 4 \end{array}$

17. $\begin{array}{r} 43 \\ 2\overline{)87} \\ 80 \\ \hline 7 \\ 6 \\ \hline 1 \end{array}$

18. $\begin{array}{r} 14 \\ 6\overline{)89} \\ 60 \\ \hline 29 \\ 24 \\ \hline 5 \end{array}$

19. $\begin{array}{r} 13 \\ 4\overline{)55} \\ 40 \\ \hline 15 \\ 12 \\ \hline 3 \end{array}$

20. $\begin{array}{r} 16 \\ 4\overline{)67} \\ 40 \\ \hline 27 \\ 24 \\ \hline 3 \end{array}$

Name: **Lesson 2-6 Dividing by one digit numbers with remainders**

1. $\begin{array}{r} 16 \\ 2\overline{)33} \\ 20 \\ \hline 13 \\ 12 \\ \hline 1 \end{array}$

2. $\begin{array}{r} 17 \\ 5\overline{)88} \\ 50 \\ \hline 38 \\ 35 \\ \hline 3 \end{array}$

3. $\begin{array}{r} 23 \\ 3\overline{)71} \\ 60 \\ \hline 11 \\ 9 \\ \hline 2 \end{array}$

4. $\begin{array}{r} 11 \\ 7\overline{)79} \\ 70 \\ \hline 9 \\ 7 \\ \hline 2 \end{array}$

5. $\begin{array}{r} 12 \\ 5\overline{)64} \\ 50 \\ \hline 14 \\ 10 \\ \hline 4 \end{array}$

6. $\begin{array}{r} 14 \\ 3\overline{)43} \\ 30 \\ \hline 13 \\ 12 \\ \hline 1 \end{array}$

7. $\begin{array}{r} 15 \\ 6\overline{)92} \\ 60 \\ \hline 32 \\ 30 \\ \hline 2 \end{array}$

8. $\begin{array}{r} 19 \\ 4\overline{)79} \\ 40 \\ \hline 39 \\ 36 \\ \hline 3 \end{array}$

9. $\begin{array}{r} 21 \\ 3\overline{)64} \\ 60 \\ \hline 4 \\ 3 \\ \hline 1 \end{array}$

10. $\begin{array}{r} 20 \\ 4\overline{)83} \\ 80 \\ \hline 3 \end{array}$

11. $\begin{array}{r} 14 \\ 6\overline{)85} \\ 60 \\ \hline 25 \\ 24 \\ \hline 1 \end{array}$

12. $\begin{array}{r} 28 \\ 2\overline{)57} \\ 40 \\ \hline 17 \\ 16 \\ \hline 1 \end{array}$

13. $\begin{array}{r} 19 \\ 4\overline{)78} \\ 40 \\ \hline 38 \\ 36 \\ \hline 2 \end{array}$

14. $\begin{array}{r} 25 \\ 3\overline{)77} \\ 60 \\ \hline 17 \\ 15 \\ \hline 2 \end{array}$

15. $\begin{array}{r} 15 \\ 6\overline{)92} \\ 60 \\ \hline 32 \\ 30 \\ \hline 2 \end{array}$

16. $\begin{array}{r} 19 \\ 3\overline{)58} \\ 30 \\ \hline 28 \\ 27 \\ \hline 1 \end{array}$

17. $\begin{array}{r} 48 \\ 2\overline{)97} \\ 80 \\ \hline 17 \\ 16 \\ \hline 1 \end{array}$

18. $\begin{array}{r} 15 \\ 6\overline{)95} \\ 60 \\ \hline 35 \\ 30 \\ \hline 5 \end{array}$

19. $\begin{array}{r} 18 \\ 3\overline{)55} \\ 30 \\ \hline 25 \\ 24 \\ \hline 1 \end{array}$

20. $\begin{array}{r} 11 \\ 6\overline{)71} \\ 60 \\ \hline 11 \\ 6 \\ \hline 5 \end{array}$

Name: Lesson 2-7 Dividing by one digit numbers with remainders

1.	21	2)43	40	3	2	1
2.	11	3)35	30	5	3	2
3.	14	6)85	60	25	24	1
4.	11	4)47	40	7	4	3
5.	10	5)51	50	1		
6.	12	6)76	60	16	12	4
7.	26	3)79	60	19	18	1
8.	13	7)92	70	22	21	1
9.	17	3)52	30	22	21	1
10.	13	4)54	40	14	12	2
11.	13	6)79	60	19	18	1
12.	11	5)58	50	8	5	3
13.	21	3)65	60	5	3	2
14.	19	5)97	50	47	45	2
15.	11	8)91	80	11	8	3
16.	12	8)99	80	19	16	3
17.	19	5)99	50	49	45	4
18.	26	3)79	60	19	18	1
19.	30	3)92	90	2		
20.	12	7)87	70	17	14	3

Name: Lesson 2-8 Dividing by one digit numbers with remainders

1.	25	2)51	40	11	10	1
2.	15	3)47	30	17	15	2
3.	14	6)89	60	29	24	5
4.	11	8)90	80	10	8	2
5.	12	5)64	50	14	10	4
6.	17	5)89	50	39	35	4
7.	27	3)83	60	23	21	2
8.	23	4)93	80	13	12	1
9.	31	2)63	60	3	2	1
10.	29	3)88	60	28	27	1
11.	12	4)50	40	10	8	2
12.	12	3)38	30	8	6	2
13.	12	6)73	60	13	12	1
14.	19	5)96	50	46	45	1
15.	16	5)84	50	34	30	4
16.	22	4)89	80	9	8	1
17.	12	4)51	40	11	8	3
18.	24	3)73	60	13	12	1
19.	22	3)67	60	7	6	1
20.	19	3)59	30	29	27	2

Name: Lesson 2-9 Dividing by one digit numbers with remainders

1.	18	3)55	30	25	24	1
2.	30	3)91	90	1		
3.	18	4)75	40	35	32	3
4.	12	6)72	60	12	12	0
5.	11	6)68	60	8	6	2
6.	12	7)90	70	20	14	6
7.	24	4)98	80	18	16	2
8.	15	5)78	50	28	25	3
9.	22	3)67	60	7	6	1
10.	11	7)79	70	9	7	2
11.	12	8)97	80	17	16	1
12.	13	5)67	50	17	15	2
13.	11	6)69	60	9	6	3
14.	31	3)95	90	5	3	2
15.	15	4)61	40	21	20	1
16.	17	5)89	50	39	35	4
17.	25	3)76	60	16	15	1
18.	18	4)75	40	35	32	3
19.	27	3)83	60	23	21	2
20.	17	3)52	30	22	21	1

Name: Lesson 2-10 Dividing by one digit numbers with remainders

1.	14	4)57	40	17	16	1
2.	22	4)90	80	10	8	2
3.	19	5)99	50	49	45	4
4.	15	4)63	40	23	20	3
5.	14	6)86	60	26	24	2
6.	23	4)94	80	14	12	2
7.	28	3)85	60	25	24	1
8.	10	9)98	90	8		
9.	12	7)89	70	19	14	5
10.	12	4)49	40	9	8	1
11.	13	7)95	70	25	21	4
12.	19	5)99	50	49	45	4
13.	37	2)75	60	15	14	1
14.	19	5)97	50	47	45	2
15.	18	5)93	50	43	40	3
16.	16	4)67	40	27	24	3
17.	16	5)81	50	31	30	1
18.	17	5)87	50	37	35	2
19.	11	7)82	70	12	7	5
20.	12	6)77	60	17	12	5

Name: **Lesson 3-1 Multiplying two and three digit numbers**

1. $\begin{array}{r} 124 \\ \times \quad 58 \\ \hline 992 \\ 620 \\ \hline 7192 \end{array}$
2. $\begin{array}{r} 285 \\ \times \quad 26 \\ \hline 1710 \\ 570 \\ \hline 7410 \end{array}$
3. $\begin{array}{r} 336 \\ \times \quad 25 \\ \hline 1680 \\ 672 \\ \hline 8400 \end{array}$
4. $\begin{array}{r} 427 \\ \times \quad 49 \\ \hline 3843 \\ 1708 \\ \hline 20923 \end{array}$
5. $\begin{array}{r} 412 \\ \times \quad 74 \\ \hline 1648 \\ 2884 \\ \hline 30488 \end{array}$
6. $\begin{array}{r} 228 \\ \times \quad 36 \\ \hline 1368 \\ 684 \\ \hline 8208 \end{array}$
7. $\begin{array}{r} 315 \\ \times \quad 56 \\ \hline 1890 \\ 1575 \\ \hline 17640 \end{array}$
8. $\begin{array}{r} 524 \\ \times \quad 53 \\ \hline 1572 \\ 2620 \\ \hline 27772 \end{array}$
9. $\begin{array}{r} 468 \\ \times \quad 43 \\ \hline 1404 \\ 1872 \\ \hline 20124 \end{array}$
10. $\begin{array}{r} 413 \\ \times \quad 37 \\ \hline 2891 \\ 1239 \\ \hline 15281 \end{array}$
11. $\begin{array}{r} 257 \\ \times \quad 46 \\ \hline 1542 \\ 1028 \\ \hline 11822 \end{array}$
12. $\begin{array}{r} 396 \\ \times \quad 29 \\ \hline 3564 \\ 792 \\ \hline 11484 \end{array}$

Name: **Lesson 3-2 Multiplying two and three digit numbers**

1. $\begin{array}{r} 337 \\ \times \quad 45 \\ \hline 1685 \\ 1348 \\ \hline 15165 \end{array}$
2. $\begin{array}{r} 513 \\ \times \quad 42 \\ \hline 1026 \\ 2052 \\ \hline 21546 \end{array}$
3. $\begin{array}{r} 535 \\ \times \quad 48 \\ \hline 4280 \\ 2140 \\ \hline 25680 \end{array}$
4. $\begin{array}{r} 442 \\ \times \quad 53 \\ \hline 1326 \\ 2210 \\ \hline 23426 \end{array}$
5. $\begin{array}{r} 542 \\ \times \quad 35 \\ \hline 2710 \\ 1626 \\ \hline 18970 \end{array}$
6. $\begin{array}{r} 299 \\ \times \quad 57 \\ \hline 2093 \\ 1495 \\ \hline 17043 \end{array}$
7. $\begin{array}{r} 583 \\ \times \quad 42 \\ \hline 1166 \\ 2332 \\ \hline 24486 \end{array}$
8. $\begin{array}{r} 458 \\ \times \quad 55 \\ \hline 2290 \\ 2290 \\ \hline 25190 \end{array}$
9. $\begin{array}{r} 319 \\ \times \quad 43 \\ \hline 957 \\ 1276 \\ \hline 13717 \end{array}$
10. $\begin{array}{r} 359 \\ \times \quad 38 \\ \hline 2872 \\ 1077 \\ \hline 13642 \end{array}$
11. $\begin{array}{r} 478 \\ \times \quad 57 \\ \hline 3346 \\ 2390 \\ \hline 27246 \end{array}$
12. $\begin{array}{r} 456 \\ \times \quad 54 \\ \hline 1824 \\ 2280 \\ \hline 24624 \end{array}$

Name: **Lesson 3-3 Multiplying two and three digit numbers**

1. $\begin{array}{r} 443 \\ \times \quad 35 \\ \hline 2215 \\ 1329 \\ \hline 15505 \end{array}$
2. $\begin{array}{r} 524 \\ \times \quad 31 \\ \hline 524 \\ 1572 \\ \hline 16244 \end{array}$
3. $\begin{array}{r} 466 \\ \times \quad 36 \\ \hline 2796 \\ 1398 \\ \hline 16776 \end{array}$
4. $\begin{array}{r} 578 \\ \times \quad 46 \\ \hline 3468 \\ 2312 \\ \hline 26588 \end{array}$
5. $\begin{array}{r} 569 \\ \times \quad 47 \\ \hline 3983 \\ 2276 \\ \hline 26743 \end{array}$
6. $\begin{array}{r} 396 \\ \times \quad 42 \\ \hline 792 \\ 1584 \\ \hline 16632 \end{array}$
7. $\begin{array}{r} 496 \\ \times \quad 37 \\ \hline 3472 \\ 1488 \\ \hline 18352 \end{array}$
8. $\begin{array}{r} 436 \\ \times \quad 33 \\ \hline 1308 \\ 1308 \\ \hline 14388 \end{array}$
9. $\begin{array}{r} 569 \\ \times \quad 49 \\ \hline 5121 \\ 2276 \\ \hline 27881 \end{array}$
10. $\begin{array}{r} 354 \\ \times \quad 42 \\ \hline 708 \\ 1416 \\ \hline 14868 \end{array}$
11. $\begin{array}{r} 451 \\ \times \quad 48 \\ \hline 3608 \\ 1804 \\ \hline 21648 \end{array}$
12. $\begin{array}{r} 341 \\ \times \quad 57 \\ \hline 2387 \\ 1705 \\ \hline 19437 \end{array}$

Name: **Lesson 3-4 Multiplying two and three digit numbers**

1. $\begin{array}{r} 426 \\ \times \quad 53 \\ \hline 1278 \\ 2130 \\ \hline 22578 \end{array}$
2. $\begin{array}{r} 465 \\ \times \quad 53 \\ \hline 1395 \\ 2325 \\ \hline 24645 \end{array}$
3. $\begin{array}{r} 563 \\ \times \quad 51 \\ \hline 563 \\ 2815 \\ \hline 28713 \end{array}$
4. $\begin{array}{r} 563 \\ \times \quad 45 \\ \hline 2815 \\ 2252 \\ \hline 25335 \end{array}$
5. $\begin{array}{r} 597 \\ \times \quad 48 \\ \hline 4776 \\ 2388 \\ \hline 28656 \end{array}$
6. $\begin{array}{r} 684 \\ \times \quad 48 \\ \hline 5472 \\ 2736 \\ \hline 32832 \end{array}$
7. $\begin{array}{r} 674 \\ \times \quad 48 \\ \hline 5392 \\ 2696 \\ \hline 32352 \end{array}$
8. $\begin{array}{r} 574 \\ \times \quad 59 \\ \hline 5166 \\ 2870 \\ \hline 33866 \end{array}$
9. $\begin{array}{r} 676 \\ \times \quad 56 \\ \hline 4056 \\ 3380 \\ \hline 37856 \end{array}$
10. $\begin{array}{r} 536 \\ \times \quad 56 \\ \hline 3216 \\ 2680 \\ \hline 30016 \end{array}$
11. $\begin{array}{r} 657 \\ \times \quad 44 \\ \hline 2628 \\ 2628 \\ \hline 28908 \end{array}$
12. $\begin{array}{r} 593 \\ \times \quad 47 \\ \hline 4151 \\ 2372 \\ \hline 27871 \end{array}$

Name: **Lesson 3-5 Multiplying two and three digit numbers**

1. 335 × 57 2345 1675 19095	2. 546 × 51 546 2730 27846	3. 547 × 36 3282 1641 19692
4. 562 × 48 4496 2248 26976	5. 429 × 44 1716 1716 18876	6. 431 × 57 3017 2155 24567
7. 578 × 39 5202 1734 22542	8. 555 × 55 2775 2775 30525	9. 347 × 46 2082 1388 15962
10. 473 × 45 2365 1892 21285	11. 469 × 47 3283 1876 22043	12. 562 × 54 2248 2810 30348

Name: **Lesson 3-6 Multiplying two and three digit numbers**

1. 386 × 52 772 1930 20072	2. 556 × 65 2780 3336 36140	3. 587 × 51 587 2935 29937
4. 637 × 45 3185 2548 28665	5. 438 × 57 3066 2190 24966	6. 473 × 42 946 1892 19866
7. 528 × 63 1584 3168 33264	8. 687 × 68 5496 4122 46716	9. 682 × 54 2728 3410 36828
10. 696 × 64 2784 4176 44544	11. 398 × 59 3582 1990 23482	12. 541 × 69 4869 3246 37329

Name: **Lesson 3-7 Multiplying two and three digit numbers**

1. 563 × 64 2252 3378 36032	2. 489 × 28 3912 978 13692	3. 667 × 48 5336 2668 32016
4. 475 × 57 3325 2375 27075	5. 558 × 64 2232 3348 35712	6. 398 × 59 3582 1990 23482
7. 663 × 49 5967 2652 32487	8. 583 × 37 4081 1749 21571	9. 568 × 68 4544 3408 38624
10. 438 × 66 2628 2628 28908	11. 643 × 57 4501 3215 36651	12. 592 × 57 4144 2960 33744

Name: **Lesson 3-8 Multiplying two and three digit numbers**

1. 654 × 74 2616 4578 48396	2. 757 × 48 6056 3028 36336	3. 723 × 52 1446 3615 37596
4. 536 × 69 4824 3216 36984	5. 468 × 73 1404 3276 34164	6. 465 × 74 1860 3255 34410
7. 437 × 72 874 3059 31464	8. 634 × 69 5706 3804 43746	9. 586 × 65 2930 3516 38090
10. 648 × 57 4536 3240 36936	11. 499 × 69 4491 2994 34431	12. 647 × 57 4529 3235 36879

Name: **Lesson 3-9** Multiplying two and three digit numbers

1.	587 × 63	1761	3522	36981
2.	577 × 56	3462	2885	32312
3.	584 × 67	4088	3504	39128
4.	833 × 56	4998	4165	46648
5.	726 × 48	5808	2904	34848
6.	746 × 54	2984	3730	40284
7.	488 × 75	2440	3416	36600
8.	657 × 69	5913	3942	45333
9.	495 × 77	3465	3465	38115
10.	597 × 66	3582	3582	39402
11.	845 × 78	6760	5915	65910
12.	834 × 85	4170	6672	70890

Name: **Lesson 3-10** Multiplying two and three digit numbers

1.	467 × 73	1401	3269	34091
2.	453 × 79	4077	3171	35787
3.	514 × 88	4112	4112	45232
4.	656 × 58	5248	3280	38048
5.	537 × 84	2148	4296	45108
6.	636 × 75	3180	4452	47700
7.	789 × 84	3156	6312	66276
8.	756 × 78	6048	5292	58968
9.	941 × 92	1882	8469	86572
10.	489 × 96	2934	4401	46944
11.	649 × 93	1947	5841	60357
12.	878 × 97	6146	7902	85166

Name: **Lesson 4-1** Dividing by one digit numbers with remainders

1.	8)322	40	320, 2
2.	4)311	77	280, 31, 28, 3
3.	5)432	86	400, 32, 30, 2
4.	7)472	67	420, 52, 49, 3
5.	6)275	45	240, 35, 30, 5
6.	7)496	70	490, 6
7.	6)590	98	540, 50, 48, 2
8.	4)311	77	280, 31, 28, 3
9.	7)456	65	420, 36, 35, 1
10.	9)245	27	180, 65, 63, 2
11.	4)337	84	320, 17, 16, 1
12.	6)298	49	240, 58, 54, 4
13.	3)404	134	300, 104, 90, 14, 12, 2
14.	6)386	64	360, 26, 24, 2
15.	9)259	28	180, 79, 72, 7
16.	4)365	91	360, 5, 4, 1

Name: **Lesson 4-2** Dividing by one digit numbers with remainders

1.	6)314	52	300, 14, 12, 2
2.	8)425	53	400, 25, 24, 1
3.	6)457	76	420, 37, 36, 1
4.	4)405	101	400, 5, 4, 1
5.	7)562	80	560, 2
6.	9)823	91	810, 13, 9, 4
7.	8)538	67	480, 58, 56, 2
8.	8)510	63	480, 30, 24, 6
9.	9)359	39	270, 89, 81, 8
10.	7)446	63	420, 26, 21, 5
11.	7)447	63	420, 27, 21, 6
12.	6)493	82	480, 13, 12, 1
13.	3)421	140	300, 121, 120, 1
14.	7)512	73	490, 22, 21, 1
15.	9)348	38	270, 78, 72, 6
16.	8)466	58	400, 66, 64, 2

Name: **Lesson 4-3 Dividing by one digit numbers with remainders**

1. 54
8)438
400
38
32
6

2. 60
7)424
420
4

3. 62
9)565
540
25
18
7

4. 104
4)418
400
18
16
2

5. 89
6)536
480
56
54
2

6. 67
6)404
360
44
42
2

7. 45
7)318
280
38
35
3

8. 90
6)543
540
3

9. 83
4)333
320
13
12
1

10. 99
3)299
270
29
27
2

11. 73
3)220
210
10
9
1

12. 55
7)387
350
37
35
2

13. 98
5)492
450
42
40
2

14. 73
7)512
490
22
21
1

15. 62
8)503
480
23
16
7

16. 62
9)564
540
24
18
6

Name: **Lesson 4-4 Dividing by one digit numbers with remainders**

1. 55
7)391
350
41
35
6

2. 76
5)384
350
34
30
4

3. 50
7)355
350
5

4. 64
4)257
240
17
16
1

5. 94
5)474
450
24
20
4

6. 166
3)500
300
200
180
20
18
2

7. 79
6)479
420
59
54
5

8. 50
8)404
400
4

9. 92
6)555
540
15
12
3

10. 63
7)443
420
23
21
2

11. 48
6)292
240
52
48
4

12. 50
6)301
300
1

13. 45
9)405
360
45
45
0

14. 99
4)398
360
38
36
2

15. 68
8)551
480
71
64
7

16. 62
9)563
540
23
18
5

Name: **Lesson 4-5 Dividing by one digit numbers with remainders**

1. 96
3)290
270
20
18
2

2. 54
4)219
200
19
16
3

3. 53
7)377
350
27
21
6

4. 75
7)526
490
36
35
1

5. 93
4)375
360
15
12
3

6. 74
7)520
490
30
28
2

7. 80
4)321
320
1

8. 92
6)555
540
15
12
3

9. 57
7)404
350
54
49
5

10. 55
9)499
450
49
45
4

11. 96
6)579
540
39
36
3

12. 52
9)470
450
20
18
2

13. 70
8)565
560
5

14. 145
3)436
300
136
120
16
15
1

15. 162
3)488
300
188
180
8
6
2

16. 54
7)383
350
33
28
5

Name: **Lesson 4-6 Dividing by one digit numbers with remainders**

1. 73
5)366
350
16
15
1

2. 137
3)412
300
112
90
22
21
1

3. 66
8)530
480
50
48
2

4. 141
3)423
300
123
120
3
3
0

5. 105
6)632
600
32
30
2

6. 79
7)556
490
66
63
3

7. 71
7)499
490
9
7
2

8. 95
7)667
630
37
35
2

9. 79
7)554
490
64
63
1

10. 41
8)333
320
13
8
5

11. 24
9)222
180
42
36
6

12. 115
4)461
400
61
40
21
20
1

13. 55
8)447
400
47
40
7

14. 73
9)664
630
34
27
7

15. 68
9)619
540
79
72
7

16. 85
7)600
560
40
35
5

Name:

Lesson 4-7 Dividing by one digit numbers with remainders

1. 85 / 6)511 / 480 / 31 / 30 / 1
2. 92 / 5)461 / 450 / 11 / 10 / 1
3. 134 / 4)537 / 400 / 137 / 120 / 17 / 16 / 1
4. 87 / 7)613 / 560 / 53 / 49 / 4
5. 86 / 7)603 / 560 / 43 / 42 / 1
6. 50 / 9)452 / 450 / 2
7. 66 / 7)466 / 420 / 46 / 42 / 4
8. 108 / 5)544 / 500 / 44 / 40 / 4
9. 57 / 8)457 / 400 / 57 / 56 / 1
10. 74 / 7)523 / 490 / 33 / 28 / 5
11. 72 / 8)577 / 560 / 17 / 16 / 1
12. 70 / 9)632 / 630 / 2
13. 77 / 9)699 / 630 / 69 / 63 / 6
14. 83 / 8)670 / 640 / 30 / 24 / 6
15. 82 / 4)330 / 320 / 10 / 8 / 2
16. 76 / 8)614 / 560 / 54 / 48 / 6

Name:

Lesson 4-8 Dividing by one digit numbers with remainders

1. 104 / 7)732 / 700 / 32 / 28 / 4
2. 94 / 7)662 / 630 / 32 / 28 / 4
3. 78 / 8)631 / 560 / 71 / 64 / 7
4. 197 / 4)789 / 400 / 389 / 360 / 29 / 28 / 1
5. 127 / 5)638 / 500 / 138 / 100 / 38 / 35 / 3
6. 57 / 9)521 / 450 / 71 / 63 / 8
7. 42 / 7)300 / 280 / 20 / 14 / 6
8. 66 / 9)599 / 540 / 59 / 54 / 5
9. 68 / 9)619 / 540 / 79 / 72 / 7
10. 71 / 6)431 / 420 / 11 / 6 / 5
11. 102 / 4)411 / 400 / 11 / 8 / 3
12. 78 / 8)631 / 560 / 71 / 64 / 7
13. 59 / 8)474 / 400 / 74 / 72 / 2
14. 61 / 9)556 / 540 / 16 / 9 / 7
15. 84 / 6)509 / 480 / 29 / 24 / 5
16. 78 / 7)551 / 490 / 61 / 56 / 5

Name:

Lesson 4-9 Dividing by one digit numbers with remainders

1. 233 / 3)700 / 600 / 100 / 90 / 10 / 9 / 1
2. 88 / 8)711 / 640 / 71 / 64 / 7
3. 67 / 9)604 / 540 / 64 / 63 / 1
4. 110 / 6)665 / 600 / 65 / 60 / 5
5. 49 / 9)446 / 360 / 86 / 81 / 5
6. 89 / 7)624 / 560 / 64 / 63 / 1
7. 89 / 6)537 / 480 / 57 / 54 / 3
8. 63 / 9)573 / 540 / 33 / 27 / 6
9. 62 / 7)437 / 420 / 17 / 14 / 3
10. 50 / 9)453 / 450 / 3
11. 159 / 4)639 / 400 / 239 / 200 / 39 / 36 / 3
12. 55 / 6)333 / 300 / 33 / 30 / 3
13. 125 / 6)752 / 600 / 152 / 120 / 32 / 30 / 2
14. 91 / 8)729 / 720 / 9 / 8 / 1
15. 85 / 9)771 / 720 / 51 / 45 / 6
16. 78 / 7)548 / 490 / 58 / 56 / 2

Name:

Lesson 4-10 Dividing by one digit numbers with remainders

1. 91 / 9)821 / 810 / 11 / 9 / 2
2. 217 / 3)652 / 600 / 52 / 30 / 22 / 21 / 1
3. 79 / 8)634 / 560 / 74 / 72 / 2
4. 82 / 7)579 / 560 / 19 / 14 / 5
5. 91 / 8)734 / 720 / 14 / 8 / 6
6. 87 / 8)700 / 640 / 60 / 56 / 4
7. 153 / 5)766 / 500 / 266 / 250 / 16 / 15 / 1
8. 81 / 9)732 / 720 / 12 / 9 / 3
9. 128 / 6)773 / 600 / 173 / 120 / 53 / 48 / 5
10. 92 / 7)646 / 630 / 16 / 14 / 2
11. 60 / 9)541 / 540 / 1
12. 77 / 8)623 / 560 / 63 / 56 / 7
13. 124 / 5)622 / 500 / 122 / 100 / 22 / 20 / 2
14. 82 / 9)746 / 720 / 26 / 18 / 8
15. 232 / 3)698 / 600 / 98 / 90 / 8 / 6 / 2
16. 142 / 4)571 / 400 / 171 / 160 / 11 / 8 / 3

Name: Lesson 5-1 Multiplying three digit numbers

1. $$\begin{array}{r} 332 \\ \times\ 416 \\ \hline 1992 \\ 332 \\ 1328 \\ \hline 138112 \end{array}$$

2. $$\begin{array}{r} 273 \\ \times\ 457 \\ \hline 1911 \\ 1365 \\ 1092 \\ \hline 124761 \end{array}$$

3. $$\begin{array}{r} 445 \\ \times\ 379 \\ \hline 4005 \\ 3115 \\ 1335 \\ \hline 168655 \end{array}$$

4. $$\begin{array}{r} 428 \\ \times\ 439 \\ \hline 3852 \\ 1284 \\ 1712 \\ \hline 187892 \end{array}$$

5. $$\begin{array}{r} 529 \\ \times\ 375 \\ \hline 2645 \\ 3703 \\ 1587 \\ \hline 198375 \end{array}$$

6. $$\begin{array}{r} 362 \\ \times\ 348 \\ \hline 2896 \\ 1448 \\ 1086 \\ \hline 125976 \end{array}$$

7. $$\begin{array}{r} 289 \\ \times\ 598 \\ \hline 2312 \\ 2601 \\ 1445 \\ \hline 172822 \end{array}$$

8. $$\begin{array}{r} 374 \\ \times\ 478 \\ \hline 2992 \\ 2618 \\ 1496 \\ \hline 178772 \end{array}$$

9. $$\begin{array}{r} 469 \\ \times\ 395 \\ \hline 2345 \\ 4221 \\ 1407 \\ \hline 185255 \end{array}$$

Name: Lesson 5-2 Multiplying three digit numbers

1. $$\begin{array}{r} 349 \\ \times\ 589 \\ \hline 3141 \\ 2792 \\ 1745 \\ \hline 205561 \end{array}$$

2. $$\begin{array}{r} 492 \\ \times\ 456 \\ \hline 2952 \\ 2460 \\ 1968 \\ \hline 224352 \end{array}$$

3. $$\begin{array}{r} 546 \\ \times\ 387 \\ \hline 3822 \\ 4368 \\ 1638 \\ \hline 211302 \end{array}$$

4. $$\begin{array}{r} 553 \\ \times\ 355 \\ \hline 2765 \\ 2765 \\ 1659 \\ \hline 196315 \end{array}$$

5. $$\begin{array}{r} 517 \\ \times\ 636 \\ \hline 3102 \\ 1551 \\ 3102 \\ \hline 328812 \end{array}$$

6. $$\begin{array}{r} 657 \\ \times\ 482 \\ \hline 1314 \\ 5256 \\ 2628 \\ \hline 316674 \end{array}$$

7. $$\begin{array}{r} 547 \\ \times\ 543 \\ \hline 1641 \\ 2188 \\ 2735 \\ \hline 297021 \end{array}$$

8. $$\begin{array}{r} 579 \\ \times\ 434 \\ \hline 2316 \\ 1737 \\ 2316 \\ \hline 251286 \end{array}$$

9. $$\begin{array}{r} 349 \\ \times\ 529 \\ \hline 3141 \\ 698 \\ 1745 \\ \hline 184621 \end{array}$$

Name: Lesson 5-3 Multiplying three digit numbers

1. $$\begin{array}{r} 442 \\ \times\ 326 \\ \hline 2652 \\ 884 \\ 1326 \\ \hline 144092 \end{array}$$

2. $$\begin{array}{r} 458 \\ \times\ 394 \\ \hline 1832 \\ 4122 \\ 1374 \\ \hline 180452 \end{array}$$

3. $$\begin{array}{r} 492 \\ \times\ 357 \\ \hline 3444 \\ 2460 \\ 1476 \\ \hline 175644 \end{array}$$

4. $$\begin{array}{r} 574 \\ \times\ 468 \\ \hline 4592 \\ 3444 \\ 2296 \\ \hline 268632 \end{array}$$

5. $$\begin{array}{r} 563 \\ \times\ 536 \\ \hline 3378 \\ 1689 \\ 2815 \\ \hline 301768 \end{array}$$

6. $$\begin{array}{r} 531 \\ \times\ 642 \\ \hline 1062 \\ 2124 \\ 3186 \\ \hline 340902 \end{array}$$

7. $$\begin{array}{r} 641 \\ \times\ 515 \\ \hline 3205 \\ 641 \\ 3205 \\ \hline 330115 \end{array}$$

8. $$\begin{array}{r} 389 \\ \times\ 398 \\ \hline 3112 \\ 3501 \\ 1167 \\ \hline 154822 \end{array}$$

9. $$\begin{array}{r} 365 \\ \times\ 497 \\ \hline 2555 \\ 3285 \\ 1460 \\ \hline 181405 \end{array}$$

Name: Lesson 5-4 Multiplying three digit numbers

1. $$\begin{array}{r} 449 \\ \times\ 644 \\ \hline 1796 \\ 1796 \\ 2694 \\ \hline 289156 \end{array}$$

2. $$\begin{array}{r} 409 \\ \times\ 499 \\ \hline 3681 \\ 3681 \\ 1636 \\ \hline 204091 \end{array}$$

3. $$\begin{array}{r} 573 \\ \times\ 649 \\ \hline 5157 \\ 2292 \\ 3438 \\ \hline 371877 \end{array}$$

4. $$\begin{array}{r} 673 \\ \times\ 455 \\ \hline 3365 \\ 3365 \\ 2692 \\ \hline 306215 \end{array}$$

5. $$\begin{array}{r} 557 \\ \times\ 587 \\ \hline 3899 \\ 4456 \\ 2785 \\ \hline 326959 \end{array}$$

6. $$\begin{array}{r} 624 \\ \times\ 525 \\ \hline 3120 \\ 1248 \\ 3120 \\ \hline 327600 \end{array}$$

7. $$\begin{array}{r} 538 \\ \times\ 432 \\ \hline 1076 \\ 1614 \\ 2152 \\ \hline 232416 \end{array}$$

8. $$\begin{array}{r} 645 \\ \times\ 564 \\ \hline 2580 \\ 3870 \\ 3225 \\ \hline 363780 \end{array}$$

9. $$\begin{array}{r} 573 \\ \times\ 697 \\ \hline 4011 \\ 5157 \\ 3438 \\ \hline 399381 \end{array}$$

Name: **Lesson 5-5** Multiplying three digit numbers

1.
```
   429
×  543
  1287
 1716
2145
232947
```
2.
```
   356
×  627
  2492
  712
2136
223212
```
3.
```
   487
×  379
  4383
 3409
1461
184573
```
4.
```
   489
×  475
  2445
 3423
1956
232275
```
5.
```
   589
×  519
  5301
  589
2945
305691
```
6.
```
   537
×  448
  4296
 2148
2148
240576
```
7.
```
   467
×  684
  1868
 3736
2802
319428
```
8.
```
   685
×  676
  4110
 4795
4110
463060
```
9.
```
   387
×  399
  3483
 3483
1161
154413
```


Name: **Lesson 5-6** Multiplying three digit numbers

1.
```
   351
×  645
  1755
 1404
2106
226395
```
2.
```
   428
×  648
  3424
 1712
2568
277344
```
3.
```
   462
×  532
   924
 1386
2310
245784
```
4.
```
   431
×  688
  3448
 3448
2586
296528
```
5.
```
   674
×  597
  4718
 6066
3370
402378
```
6.
```
   667
×  344
  2668
 2668
2001
229448
```
7.
```
   507
×  417
  3549
  507
2028
211419
```
8.
```
   645
×  352
  1290
 3225
1935
227040
```
9.
```
   424
×  535
  2120
 1272
2120
226840
```


Name: **Lesson 5-7** Multiplying three digit numbers

1.
```
   374
×  575
  1870
 2618
1870
215050
```
2.
```
   587
×  696
  3522
 5283
3522
408552
```
3.
```
   597
×  593
  1791
 5373
2985
354021
```
4.
```
   684
×  784
  2736
 5472
4788
536256
```
5.
```
   758
×  493
  2274
 6822
3032
373694
```
6.
```
   627
×  733
  1881
 1881
4389
459591
```
7.
```
   776
×  376
  4656
 5432
2328
291776
```
8.
```
   591
×  724
  2364
 1182
4137
427884
```
9.
```
   468
×  468
  3744
 2808
1872
219024
```


Name: **Lesson 5-8** Multiplying three digit numbers

1.
```
   456
×  579
  4104
 3192
2280
264024
```
2.
```
   497
×  486
  2982
 3976
1988
241542
```
3.
```
   452
×  499
  4068
 4068
1808
225548
```
4.
```
   658
×  452
  1316
 3290
2632
297416
```
5.
```
   753
×  375
  3765
 5271
2259
282375
```
6.
```
   683
×  615
  3415
  683
4098
420045
```
7.
```
   741
×  489
  6669
 5928
2964
362349
```
8.
```
   636
×  531
   636
 1908
3180
337716
```
9.
```
   493
×  793
  1479
 4437
3451
390949
```

Name:

Lesson 5-9 Multiplying three digit numbers

1.
484
× 776
2904
3388
3388
375584

2.
557
× 586
3342
4456
2785
326402

3.
488
× 568
3904
2928
2440
277184

4.
658
× 547
4606
2632
3290
359926

5.
717
× 526
4302
1434
3585
377142

6.
674
× 698
5392
6066
4044
470452

7.
773
× 577
5411
5411
3865
446021

8.
743
× 357
5201
3715
2229
265251

9.
575
× 683
1725
4600
3450
392725

Name:

Lesson 5-10 Multiplying three digit numbers

1.
589
× 677
4123
4123
3534
398753

2.
565
× 487
3955
4520
2260
275155

3.
465
× 564
1860
2790
2325
262260

4.
777
× 454
3108
3885
3108
352758

5.
783
× 409
7047
3132 ·
320247

6.
784
× 678
6272
5488
4704
531552

7.
687
× 642
1374
2748
4122
441054

8.
834
× 513
2502
834
4170
427842

9.
775
× 567
5425
4650
3875
439425

Name:

Lesson 6-1 Dividing by two digit numbers with remainders

1.
20
32) 654
640
14

2.
15
37) 572
370
202
185
17

3.
18
31) 576
310
266
248
18

4.
16
34) 545
340
205
204
1

5.
19
25) 498
250
248
225
23

6.
14
46) 661
460
201
184
17

7.
17
27) 468
270
198
189
9

8.
18
26) 488
260
228
208
20

9.
20
17) 351
340
11

10.
15
38) 595
380
215
190
25

11.
14
39) 549
390
159
156
3

12.
24
19) 461
380
81
76
5

13.
29
19) 565
380
185
171
14

14.
22
23) 512
460
52
46
6

15.
28
17) 478
340
138
136
2

16.
12
37) 475
370
105
74
31

Name:

Lesson 6-2 Dividing by two digit numbers with remainders

1.
13
43) 592
430
162
129
33

2.
51
15) 778
750
28
15
13

3.
40
16) 653
640
13

4.
28
27) 774
540
234
216
18

5.
20
31) 624
620
4

6.
24
27) 660
540
120
108
12

7.
21
28) 598
560
38
28
10

8.
19
35) 688
350
338
315
23

9.
13
37) 505
370
135
111
24

10.
22
29) 655
580
75
58
17

11.
19
31) 600
310
290
279
11

12.
35
17) 611
510
101
85
16

13.
15
42) 633
420
213
210
3

14.
28
26) 735
520
215
208
7

15.
19
29) 577
290
287
261
26

16.
28
21) 594
420
174
168
6

```
        22
  34 ) 753
       680
        73
        68
         5
```
2.
```
        28
  24 ) 679
       480
       199
       192
         7
```
3.
```
        21
  35 ) 766
       700
        66
        35
        31
```
4.
```
        23
  31 ) 735
       620
       115
        93
        22
```

5.
```
        16
  41 ) 659
       410
       249
       246
         3
```
6.
```
        48
  16 ) 770
       640
       130
       128
         2
```
7.
```
        37
  17 ) 635
       510
       125
       119
         6
```
8.
```
        35
  18 ) 639
       540
        99
        90
         9
```

9.
```
        35
  15 ) 531
       450
        81
        75
         6
```
10.
```
        23
  26 ) 622
       520
       102
        78
        24
```
11.
```
        58
  13 ) 757
       650
       107
       104
         3
```
12.
```
        36
  19 ) 686
       570
       116
       114
         2
```

13.
```
        49
  16 ) 790
       640
       150
       144
         6
```
14.
```
        40
  17 ) 696
       680
        16
```
15.
```
        15
  37 ) 564
       370
       194
       185
         9
```
16.
```
        21
  26 ) 547
       520
        27
        26
         1
```

1.
```
        20
  41 ) 831
       820
        11
```
2.
```
        22
  34 ) 767
       680
        87
        68
        19
```
3.
```
        14
  42 ) 589
       420
       169
       168
         1
```
4.
```
        13
  63 ) 844
       630
       214
       189
        25
```

5.
```
        31
  24 ) 767
       720
        47
        24
        23
```
6.
```
        45
  19 ) 864
       760
       104
        95
         9
```
7.
```
        52
  16 ) 838
       800
        38
        32
         6
```
8.
```
        31
  23 ) 715
       690
        25
        23
         2
```

9.
```
        38
  17 ) 662
       510
       152
       136
        16
```
10.
```
        14
  53 ) 743
       530
       213
       212
         1
```
11.
```
        12
  56 ) 726
       560
       166
       112
        54
```
12.
```
        63
  12 ) 767
       720
        47
        36
        11
```

13.
```
        22
  35 ) 786
       700
        86
        70
        16
```
14.
```
        39
  17 ) 678
       510
       168
       153
        15
```
15.
```
        49
  12 ) 599
       480
       119
       108
        11
```
16.
```
        29
  21 ) 620
       420
       200
       189
        11
```

1.
```
        42
  15 ) 632
       600
        32
        30
         2
```
2.
```
        31
  14 ) 439
       420
        19
        14
         5
```
3.
```
        17
  35 ) 599
       350
       249
       245
         4
```
4.
```
        31
  24 ) 754
       720
        34
        24
        10
```

5.
```
        38
  20 ) 765
       600
       165
       160
         5
```
6.
```
        20
  27 ) 566
       540
        26
```
7.
```
        45
  11 ) 497
       440
        57
        55
         2
```
8.
```
        48
  13 ) 633
       520
       113
       104
         9
```

9.
```
        45
  17 ) 776
       680
        96
        85
        11
```
10.
```
        17
  43 ) 736
       430
       306
       301
         5
```
11.
```
        32
  24 ) 783
       720
        63
        48
        15
```
12.
```
        24
  28 ) 675
       560
       115
       112
         3
```

13.
```
        18
  35 ) 647
       350
       297
       280
        17
```
14.
```
        17
  39 ) 681
       390
       291
       273
        18
```
15.
```
        13
  43 ) 595
       430
       165
       129
        36
```
16.
```
        47
  15 ) 714
       600
       114
       105
         9
```

1.
```
        29
  22 ) 641
       440
       201
       198
         3
```
2.
```
        21
  36 ) 772
       720
        52
        36
        16
```
3.
```
        13
  52 ) 714
       520
       194
       156
        38
```
4.
```
        14
  54 ) 781
       540
       241
       216
        25
```

5.
```
        13
  45 ) 589
       450
       139
       135
         4
```
6.
```
        28
  23 ) 654
       460
       194
       184
        10
```
7.
```
        27
  21 ) 582
       420
       162
       147
        15
```
8.
```
        19
  33 ) 655
       330
       325
       297
        28
```

9.
```
        65
  12 ) 785
       720
        65
        60
         5
```
10.
```
        65
  12 ) 789
       720
        69
        60
         9
```
11.
```
        39
  16 ) 627
       480
       147
       144
         3
```
12.
```
        72
  11 ) 798
       770
        28
        22
         6
```

13.
```
        17
  37 ) 644
       370
       274
       259
        15
```
14.
```
        25
  31 ) 780
       620
       160
       155
         5
```
15.
```
        18
  39 ) 737
       390
       347
       312
        35
```
16.
```
        23
  29 ) 692
       580
       112
        87
```

Name:

Lesson 6-7 Dividing by two digit numbers with remainders

1.	21	41)891	820	71	41	30
2.	14	47)682	470	212	188	24
3.	24	26)641	520	121	104	17
4.	15	45)698	450	248	225	23
5.	39	12)475	360	115	108	7
6.	33	24)793	720	73	72	1
7.	30	25)757	750	7		
8.	15	52)781	520	261	260	1
9.	21	34)743	680	63	34	29
10.	61	13)804	780	24	13	11
11.	45	19)864	760	104	95	9
12.	65	13)852	780	72	65	7
13.	50	16)813	800	13		
14.	26	27)710	540	170	162	8
15.	24	33)793	660	133	132	1
16.	31	25)785	750	35	25	10

Name:

Lesson 6-8 Dividing by two digit numbers with remainders

1.	22	25)555	500	55	50	5
2.	15	34)522	340	182	170	12
3.	33	21)697	630	67	63	4
4.	16	45)739	450	289	270	19
5.	71	11)790	770	20	11	9
6.	35	23)817	690	127	115	12
7.	16	33)546	330	216	198	18
8.	26	32)863	640	223	192	31
9.	19	41)812	410	402	369	33
10.	55	16)886	800	86	80	6
11.	55	16)887	800	87	80	7
12.	27	24)650	480	170	168	2
13.	13	63)836	630	206	189	17
14.	23	32)737	640	97	96	1
15.	26	27)705	540	165	162	3
16.	44	16)707	640	67	64	3

Name:

Lesson 6-9 Dividing by two digit numbers with remainders

1.	31	24)745	720	25	24	1
2.	33	29)964	870	94	87	7
3.	12	68)840	680	160	136	24
4.	26	32)832	640	192	192	0
5.	29	29)842	580	262	261	1
6.	16	45)747	450	297	270	27
7.	15	45)703	450	253	225	28
8.	41	17)713	680	33	17	16
9.	21	37)779	740	39	37	2
10.	15	34)518	340	178	170	8
11.	37	25)942	750	192	175	17
12.	15	41)622	410	212	205	7
13.	47	18)862	720	142	126	16
14.	76	13)999	910	89	78	11
15.	35	14)493	420	73	70	3
16.	34	28)961	840	121	112	9

Name:

Lesson 6-10 Dividing by two digit numbers with remainders

1.	22	33)753	660	93	66	27
2.	57	17)982	850	132	119	13
3.	72	12)866	840	26	24	2
4.	17	54)953	540	413	378	35
5.	29	26)756	520	236	234	2
6.	28	28)795	560	235	224	11
7.	50	17)865	850	15		
8.	42	19)802	760	42	38	4
9.	19	31)610	310	300	279	21
10.	26	32)840	640	200	192	8
11.	20	28)580	560	20		
12.	21	37)789	740	49	37	12
13.	46	18)834	720	114	108	6
14.	59	11)652	550	102	99	3
15.	14	49)722	490	232	196	36
16.	34	25)855	750	105	100	5

Name: Lesson 7-1 Multiplying three digit numbers

1.
```
     412
  ×  497
    2884
   3708
  1648
  204764
```
2.
```
     439
  ×  572
     878
   3073
  2195
  251108
```
3.
```
     524
  ×  479
    4716
   3668
  2096
  250996
```
4.
```
     571
  ×  589
    5139
   4568
  2855
  336319
```
5.
```
     681
  ×  476
    4086
   4767
  2724
  324156
```
6.
```
     297
  ×  548
    2376
   1188
  1485
  162756
```
7.
```
     728
  ×  676
    4368
   5096
  4368
  492128
```
8.
```
     683
  ×  649
    6147
   2732
  4098
  443267
```
9.
```
     649
  ×  567
    4543
   3894
  3245
  367983
```


Name: Lesson 7-2 Multiplying three digit numbers

1.
```
     577
  ×  632
    1154
   1731
  3462
  364664
```
2.
```
     357
  ×  542
     714
   1428
  1785
  193494
```
3.
```
     675
  ×  294
    2700
   6075
  1350
  198450
```
4.
```
     546
  ×  759
    4914
   2730
  3822
  414414
```
5.
```
     641
  ×  682
    1282
   5128
  3846
  437162
```
6.
```
     593
  ×  745
    2965
   2372
  4151
  441785
```
7.
```
     845
  ×  678
    6760
   5915
  5070
  572910
```
8.
```
     734
  ×  457
    5138
   3670
  2936
  335438
```
9.
```
     842
  ×  297
    5894
   7578
  1684
  250074
```


Name: Lesson 7-3 Multiplying three digit numbers

1.
```
     424
  ×  478
    3392
   2968
  1696
  202672
```
2.
```
     682
  ×  387
    4774
   5456
  2046
  263934
```
3.
```
     553
  ×  629
    4977
   1106
  3318
  347837
```
4.
```
     653
  ×  365
    3265
   3918
  1959
  238345
```
5.
```
     561
  ×  396
    3366
   5049
  1683
  222156
```
6.
```
     578
  ×  724
    2312
   1156
  4046
  418472
```
7.
```
     742
  ×  687
    5194
   5936
  4452
  509754
```
8.
```
     498
  ×  765
    2490
   2988
  3486
  380970
```
9.
```
     625
  ×  794
    2500
   5625
  4375
  496250
```


Name: Lesson 7-4 Multiplying three digit numbers

1.
```
     429
  ×  784
    1716
   3432
  3003
  336336
```
2.
```
     475
  ×  488
    3800
   3800
  1900
  231800
```
3.
```
     458
  ×  652
     916
   2290
  2748
  298616
```
4.
```
     892
  ×  315
    4460
    892
  2676
  280980
```
5.
```
     964
  ×  479
    8676
   6748
  3856
  461756
```
6.
```
     935
  ×  678
    7480
   6545
  5610
  633930
```
7.
```
     538
  ×  676
    3228
   3766
  3228
  363688
```
8.
```
     852
  ×  472
    1704
   5964
  3408
  402144
```
9.
```
     499
  ×  366
    2994
   2994
  1497
  182634
```

Name: Lesson 7-5 Multiplying three digit numbers

1.
```
    392
  × 478
   3136
  2744
 1568
 187376
```

2.
```
    489
  × 564
   1956
  2934
 2445
 275796
```

3.
```
    742
  × 648
   5936
  2968
 4452
 480816
```

4.
```
    756
  × 843
   2268
  3024
 6048
 637308
```

5.
```
    667
  × 733
   2001
  2001
 4669
 488911
```

6.
```
    537
  × 559
   4833
  2685
 2685
 300183
```

7.
```
    556
  × 696
   3336
  5004
 3336
 386976
```

8.
```
    598
  × 346
   3588
  2392
 1794
 206908
```

9.
```
    728
  × 736
   4368
  2184
 5096
 535808
```


Name: Lesson 7-6 Multiplying three digit numbers

1.
```
    256
  × 715
   1280
   256
 1792
 183040
```

2.
```
    456
  × 789
   4104
  3648
 3192
 359784
```

3.
```
    575
  × 636
   3450
  1725
 3450
 365700
```

4.
```
    563
  × 946
   3378
  2252
 5067
 532598
```

5.
```
    723
  × 478
   5784
  5061
 2892
 345594
```

6.
```
    696
  × 799
   6264
  6264
 4872
 556104
```

7.
```
    835
  × 875
   4175
  5845
 6680
 730625
```

8.
```
    592
  × 282
   1184
  4736
 1184
 166944
```

9.
```
    537
  × 655
   2685
  2685
 3222
 351735
```


Name: Lesson 7-7 Multiplying three digit numbers

1.
```
    466
  × 674
   1864
  3262
 2796
 314084
```

2.
```
    743
  × 365
   3715
  4458
 2229
 271195
```

3.
```
    894
  × 745
   4470
  3576
 6258
 666030
```

4.
```
    695
  × 658
   5560
  3475
 4170
 457310
```

5.
```
    459
  × 424
   1836
   918
 1836
 194616
```

6.
```
    783
  × 594
   3132
  7047
 3915
 465102
```

7.
```
    758
  × 549
   6822
  3032
 3790
 416142
```

8.
```
    647
  × 683
   1941
  5176
 3882
 441901
```

9.
```
    672
  × 521
    672
  1344
 3360
 350112
```


Name: Lesson 7-8 Multiplying three digit numbers

1.
```
    324
  × 487
   2268
  2592
 1296
 157788
```

2.
```
    687
  × 758
   5496
  3435
 4809
 520746
```

3.
```
    346
  × 382
    692
  2768
 1038
 132172
```

4.
```
    458
  × 976
   2748
  3206
 4122
 447008
```

5.
```
    865
  × 973
   2595
  6055
 7785
 841645
```

6.
```
    679
  × 592
   1358
  6111
 3395
 401968
```

7.
```
    765
  × 876
   4590
  5355
 6120
 670140
```

8.
```
    445
  × 871
    445
  3115
 3560
 387595
```

9.
```
    585
  × 864
   2340
  3510
 4680
 505440
```

Name: Lesson 7-9 Multiplying three digit numbers

1.	2.	3.	4.	5.	6.	7.	8.	9.
726	734	902	458	589	664	776	599	879
× 973	× 568	× 324	× 912	× 635	× 727	× 878	× 399	× 847
2178	5872	3608	916	2945	4648	6208	5391	6153
5082	4404	1804	458	1767	1328	5432	5391	3516
6534	3670	2706	4122	3534	4648	6208	1797	7032
706398	416912	292248	417696	374015	482728	681328	239001	744513

Name: Lesson 7-10 Multiplying three digit numbers

1.	2.	3.	4.	5.	6.	7.	8.	9.
849	367	645	964	609	971	875	517	556
× 634	× 498	× 862	× 969	× 369	× 538	× 385	× 628	× 975
3396	2936	1290	8676	5481	7768	4375	4136	2780
2547	3303	3870	5784	3654	2913	7000	1034	3892
5094	1468	5160	8676	1827	4855	2625	3102	5004
538266	182766	555990	934116	224721	522398	336875	324676	542100

Name: Lesson 8-1 Dividing by two digit numbers with remainders

1.	2.	3.	4.	5.	6.	7.	8.
19	13	43	23	21	30	13	39
35)672	64)861	12)523	27)645	11)239	26)790	56)773	25)992
350	640	480	540	220	780	560	750
322	221	43	105	19	10	213	242
315	192	36	81	11		168	225
7	29	7	24	8		45	17

9.	10.	11.	12.	13.	14.	15.	16.
10	13	11	29	27	16	20	20
42)434	29)389	54)632	27)784	13)357	31)512	42)845	32)647
420	290	540	540	260	310	840	640
14	99	92	244	97	202	5	7
	87	54	243	91	186		
	12	38	1	6	16		

Name: Lesson 8-2 Dividing by two digit numbers with remainders

1.	2.	3.	4.	5.	6.	7.	8.
15	17	41	35	11	32	12	19
58)881	37)650	12)495	22)784	41)452	24)779	54)680	49)941
580	370	480	660	410	720	540	490
301	280	15	124	42	59	140	451
290	259	12	110	41	48	108	441
11	21	3	14	1	11	32	10

9.	10.	11.	12.	13.	14.	15.	16.
28	71	11	10	70	13	33	51
25)707	14)997	52)587	63)633	13)919	43)584	23)771	17)872
500	980	520	630	910	430	690	850
207	17	67	3	9	154	81	22
200	14	52			129	69	17
7	3	15			25	12	5

Name: **Lesson 8-3 Dividing by two digit numbers with remainders**

1. 16 — 29)471 — 290 — 181 — 174 — 7
2. 20 — 37)750 — 740 — 10
3. 13 — 42)567 — 420 — 147 — 126 — 21
4. 55 — 15)832 — 750 — 82 — 75 — 7
5. 20 — 36)752 — 720 — 32
6. 34 — 28)960 — 840 — 120 — 112 — 8
7. 11 — 73)810 — 730 — 80 — 73 — 7
8. 10 — 56)566 — 560 — 6
9. 57 — 17)976 — 850 — 126 — 119 — 7
10. 25 — 21)530 — 420 — 110 — 105 — 5
11. 36 — 22)795 — 660 — 135 — 132 — 3
12. 30 — 24)731 — 720 — 11
13. 13 — 43)592 — 430 — 162 — 129 — 33
14. 45 — 17)777 — 680 — 97 — 85 — 12
15. 55 — 15)836 — 750 — 86 — 75 — 11
16. 18 — 48)899 — 480 — 419 — 384 — 35

Name: **Lesson 8-4 Dividing by two digit numbers with remainders**

1. 43 — 17)742 — 680 — 62 — 51 — 11
2. 33 — 26)875 — 780 — 95 — 78 — 17
3. 18 — 47)892 — 470 — 422 — 376 — 46
4. 42 — 14)594 — 560 — 34 — 28 — 6
5. 26 — 25)660 — 500 — 160 — 150 — 10
6. 32 — 14)461 — 420 — 41 — 28 — 13
7. 10 — 53)542 — 530 — 12
8. 39 — 16)633 — 480 — 153 — 144 — 9
9. 20 — 38)761 — 760 — 1
10. 18 — 39)723 — 390 — 333 — 312 — 21
11. 26 — 27)725 — 540 — 185 — 162 — 23
12. 32 — 23)743 — 690 — 53 — 46 — 7
13. 72 — 13)948 — 910 — 38 — 26 — 12
14. 14 — 67)945 — 670 — 275 — 268 — 7
15. 17 — 58)987 — 580 — 407 — 406 — 1
16. 26 — 32)837 — 640 — 197 — 192 — 5

Name: **Lesson 8-5 Dividing by two digit numbers with remainders**

1. 17 — 42)734 — 420 — 314 — 294 — 20
2. 16 — 34)561 — 340 — 221 — 204 — 17
3. 14 — 47)680 — 470 — 210 — 188 — 22
4. 30 — 31)955 — 930 — 25
5. 15 — 57)892 — 570 — 322 — 285 — 37
6. 29 — 19)556 — 380 — 176 — 171 — 5
7. 35 — 22)774 — 660 — 114 — 110 — 4
8. 40 — 16)654 — 640 — 14
9. 18 — 39)740 — 390 — 350 — 312 — 38
10. 14 — 44)649 — 440 — 209 — 176 — 33
11. 22 — 33)748 — 660 — 88 — 66 — 22
12. 35 — 22)784 — 660 — 124 — 110 — 14
13. 23 — 27)635 — 540 — 95 — 81 — 14
14. 30 — 29)888 — 870 — 18
15. 36 — 27)976 — 810 — 166 — 162 — 4
16. 23 — 42)972 — 840 — 132 — 126 — 6

Name: **Lesson 8-6 Dividing by two digit numbers with remainders**

1. 19 — 24)467 — 240 — 227 — 216 — 11
2. 16 — 54)894 — 540 — 354 — 324 — 30
3. 32 — 31)999 — 930 — 69 — 62 — 7
4. 39 — 14)551 — 420 — 131 — 126 — 5
5. 15 — 55)844 — 550 — 294 — 275 — 19
6. 30 — 22)671 — 660 — 11
7. 28 — 25)702 — 500 — 202 — 200 — 2
8. 27 — 33)922 — 660 — 262 — 231 — 31
9. 31 — 17)532 — 510 — 22 — 17 — 5
10. 25 — 37)956 — 740 — 216 — 185 — 31
11. 17 — 47)818 — 470 — 348 — 329 — 19
12. 31 — 26)827 — 780 — 47 — 26 — 21
13. 25 — 39)998 — 780 — 218 — 195 — 23
14. 18 — 41)749 — 410 — 339 — 328 — 11
15. 35 — 27)954 — 810 — 144 — 135 — 9
16. 15 — 65)997 — 650 — 347 — 325 — 22

Name: **Lesson 8-7** Dividing by two digit numbers with remainders

No.	Division	Quotient	Working
1.	44) 891	20	880, 11
2.	41) 785	19	410, 375, 369, 6
3.	63) 651	10	630, 21
4.	26) 930	35	780, 150, 130, 20
5.	31) 838	27	620, 218, 217, 1
6.	26) 587	22	520, 67, 52, 15
7.	15) 520	34	450, 70, 60, 10
8.	46) 871	18	460, 411, 368, 43
9.	12) 393	32	360, 33, 24, 9
10.	13) 636	48	520, 116, 104, 12
11.	29) 596	20	580, 16
12.	35) 952	27	700, 252, 245, 7
13.	27) 712	26	540, 172, 162, 10
14.	16) 904	56	800, 104, 96, 8
15.	44) 699	15	440, 259, 220, 39
16.	11) 478	43	440, 38, 33, 5

Name: **Lesson 8-8** Dividing by two digit numbers with remainders

No.	Division	Quotient	Working
1.	34) 555	16	340, 215, 204, 11
2.	29) 542	18	290, 252, 232, 20
3.	21) 983	46	840, 143, 126, 17
4.	28) 958	34	840, 118, 112, 6
5.	28) 709	25	560, 149, 140, 9
6.	14) 841	60	840, 1
7.	32) 940	29	640, 300, 288, 12
8.	29) 798	27	580, 218, 203, 15
9.	39) 526	13	390, 136, 117, 19
10.	16) 737	46	640, 97, 96, 1
11.	37) 760	20	740, 20
12.	38) 575	15	380, 195, 190, 5
13.	57) 683	11	570, 113, 57, 56
14.	37) 958	25	740, 218, 185, 33
15.	46) 825	17	460, 365, 322, 43
16.	18) 961	53	900, 61, 54, 7

Name: **Lesson 8-9** Dividing by two digit numbers with remainders

No.	Division	Quotient	Working
1.	42) 753	17	420, 333, 294, 39
2.	57) 976	17	570, 406, 399, 7
3.	48) 840	17	480, 360, 336, 24
4.	24) 854	35	720, 134, 120, 14
5.	26) 671	25	520, 151, 130, 21
6.	22) 853	38	660, 193, 176, 17
7.	19) 580	30	570, 10
8.	47) 669	14	470, 199, 188, 11
9.	38) 555	14	380, 175, 152, 23
10.	33) 824	24	660, 164, 132, 32
11.	23) 652	28	460, 192, 184, 8
12.	12) 473	39	360, 113, 108, 5
13.	13) 772	59	650, 122, 117, 5
14.	14) 981	70	980, 1
15.	32) 875	27	640, 235, 224, 11
16.	41) 787	19	410, 377, 369, 8

Name: **Lesson 8-10** Dividing by two digit numbers with remainders

No.	Division	Quotient	Working
1.	17) 482	28	340, 142, 136, 6
2.	39) 872	22	780, 92, 78, 14
3.	17) 488	28	340, 148, 136, 12
4.	54) 974	18	540, 434, 432, 2
5.	32) 886	27	640, 246, 224, 22
6.	22) 589	26	440, 149, 132, 17
7.	29) 893	30	870, 23
8.	16) 990	61	960, 30, 16, 14
9.	13) 310	23	260, 50, 39, 11
10.	31) 820	26	620, 200, 186, 14
11.	31) 850	27	620, 230, 217, 13
12.	37) 705	19	370, 335, 333, 2
13.	27) 832	30	810, 22
14.	42) 955	22	840, 115, 84, 31
15.	26) 735	28	520, 215, 208, 7
16.	21) 689	32	630, 59, 42, 17

Name: Lesson 9-1 Order of operations

1. $13 + 5 - 7 = 11$
2. $6 \times 9 - 18 = 36$
3. $24 - 3 \times 6 = 6$
4. $28 \div 4 + 13 = 20$
5. $24 \div 3 + 8 = 16$
6. $(23 - 5) \times 3 = 54$
7. $5 \times (25 - 6) = 95$
8. $72 \div (23 - 11) = 6$
9. $32 + 3 \times 4 = 44$
10. $55 \div (4 + 7) = 5$
11. $36 \div 2 - 9 = 9$
12. $15 - 81 \div 9 = 6$
13. $(7 + 8) \times 4 = 60$
14. $3 \times (4 + 7) = 33$
15. $9 + 30 \div 5 = 15$
16. $44 - 17 - 21 = 6$
17. $54 \div 9 \times 6 = 36$
18. $8 \times 5 + 27 = 67$
19. $64 \div 8 \div 2 = 4$
20. $16 \times 4 \div 8 = 8$
21. $(24 + 12) \div 9 = 4$
22. $4 \times 5 \times 6 = 120$
23. $(54 \div 6) \times 9 = 81$
24. $(23 - 5) \div 6 = 3$

Name: Lesson 9-2 Order of operations

1. $17 - 9 + 8 = 16$
2. $40 \div 8 + 7 = 12$
3. $33 - 2 \times 9 = 15$
4. $5 \times 9 - 8 = 37$
5. $7 \times (11 - 3) = 56$
6. $84 \div (5 + 7) = 7$
7. $25 + 11 + 9 = 45$
8. $(27 - 13) \times 8 = 112$
9. $45 + 3 \times 9 = 72$
10. $5 \times (7 + 4) = 55$
11. $6 \times (16 + 7) = 138$
12. $37 - (64 \div 4) = 21$
13. $(7 + 8) \times 4 = 60$
14. $65 - 23 - 19 = 23$
15. $90 \div (15 - 9) = 15$
16. $42 \div 7 \times 8 = 48$
17. $56 \div 8 \times 7 = 49$
18. $4 \times 8 + 15 = 47$
19. $13 + 35 \div 7 = 18$
20. $55 - 18 + 23 = 60$
21. $(29 + 27) \div 7 = 8$
22. $(32 - 7) \div 5 = 5$
23. $99 \div 3 \div 11 = 3$
24. $4 \times 3 \times 9 = 108$

Name: Lesson 9-3 Order of operations

1. $28 + 8 - 7 = 29$
2. $5 \times 7 - 18 = 17$
3. $67 - 9 \times 6 = 13$
4. $72 \div 6 + 15 = 27$
5. $14 + 27 - 13 = 28$
6. $(54 - 38) \times 4 = 64$
7. $6 \times (21 - 7) = 84$
8. $60 \div (6 + 4) = 6$
9. $33 + 4 \times 8 = 65$
10. $(42 - 18) \div 4 = 6$
11. $81 \div 9 - 6 = 3$
12. $8 \times (11 + 7) = 144$
13. $(7 + 9) \times 9 = 144$
14. $65 - 26 - 24 = 15$
15. $96 \div (12 - 4) = 12$
16. $7 \times 9 + 12 = 75$
17. $16 + 84 \div 4 = 37$
18. $6 \times 8 \div 4 = 12$
19. $42 \div 3 \times 4 = 56$
20. $56 - 24 - 17 = 15$
21. $70 \div 5 \div 2 = 7$
22. $3 \times 5 \times 8 = 120$
23. $(20 + 16) \div 2 = 18$
24. $(73 - 7) \div 11 = 6$

Name: Lesson 9-4 Order of operations

1. $37 + 11 - 25 = 23$
2. $54 \div 2 + 27 = 54$
3. $57 - 8 \times 7 = 1$
4. $7 \times 9 - 17 = 46$
5. $8 \times (13 - 7) = 48$
6. $66 \div (6 + 5) = 6$
7. $35 + 15 + 25 = 75$
8. $(43 - 26) \times 6 = 102$
9. $4 \times (21 - 12) = 36$
10. $3 \times (24 + 35) = 177$
11. $48 \div 2 - 17 = 7$
12. $48 - 16 \div 4 = 42$
13. $13 + 7 \times 7 = 62$
14. $100 - 35 - 47 = 18$
15. $(5 + 9) \times 7 = 98$
16. $13 \times 9 \div 3 = 39$
17. $98 \div (36 - 22) = 7$
18. $7 \times 3 + 59 = 80$
19. $65 \div 13 \times 12 = 60$
20. $4 \times 5 \times 7 = 140$
21. $(46 + 62) \div 9 = 12$
22. $76 - 28 + 29 = 77$
23. $126 \div 3 \div 3 = 14$
24. $(60 - 12) \div 8 = 6$

Name: Lesson 9-5 Order of operations

1. $37 + 16 - 23 = 30$
2. $7 \times 7 - 18 = 31$
3. $43 - 8 \times 4 = 11$
4. $54 \div 6 - 3 = 6$
5. $11 + 22 + 33 = 66$
6. $(31 - 13) \times 5 = 90$
7. $6 \times (21 - 8) = 78$
8. $48 - 48 \div 8 = 42$
9. $7 + 7 \times 8 = 63$
10. $36 \div (9 + 9) = 2$
11. $69 \div 3 - 4 = 19$
12. $67 - 28 - 17 = 22$
13. $(31 + 42) \times 3 = 219$
14. $5 \times (21 + 28) = 245$
15. $125 \div 5 - 8 = 17$
16. $11 \times 9 + 8 = 107$
17. $45 \div (12 - 7) = 9$
18. $45 - 21 + 14 = 38$
19. $65 \div 5 \times 8 = 104$
20. $25 \times 12 \div 3 = 100$
21. $(24 + 36) \div 4 = 15$
22. $4 \times 6 \times 6 = 144$
23. $98 \div 7 \div 2 = 7$
24. $(63 - 9) \div 2 = 27$

Name: Lesson 9-6 Order of operations

1. $26 + 15 - 13 = 28$
2. $8 \times 4 - 19 = 13$
3. $50 - (4 \times 8) = 18$
4. $72 \div 2 + 27 = 63$
5. $31 + 17 + 24 = 72$
6. $(25 - 8) \times 7 = 119$
7. $7 \times (33 - 18) = 105$
8. $160 \div (8 + 2) = 16$
9. $70 \div (14 - 9) = 14$
10. $30 - (30 \div 6) = 25$
11. $25 + 3 \times 8 = 49$
12. $7 \times (12 + 14) = 182$
13. $96 \div (22 - 6) = 6$
14. $123 - 32 - 44 = 47$
15. $(12 + 13) \times 6 = 150$
16. $7 \times 9 + 36 = 99$
17. $56 \div 8 \times 9 = 63$
18. $11 \times 12 \div 12 = 11$
19. $9 + 78 \div 13 = 15$
20. $73 - 15 + 27 = 85$
21. $153 \div 3 \div 3 = 17$
22. $5 \times 6 \times 7 = 210$
23. $(37 + 73) \div 5 = 20$
24. $(55 - 19) \div 6 = 6$

Name: Lesson 9-7 Order of operations

1. $28 + 14 - 17 = 25$
2. $6 \times 7 - 27 = 15$
3. $80 - 7 \times 8 = 24$
4. $81 \div 9 + 18 = 27$
5. $24 + 35 + 46 = 105$
6. $(31 - 19) \times 5 = 60$
7. $6 \times (32 - 17) = 90$
8. $162 \div (6 + 3) = 18$
9. $45 + 8 \times 3 = 69$
10. $35 - 63 \div 7 = 26$
11. $120 \div (32 - 8) = 5$
12. $10 \times (17 + 19) = 360$
13. $60 \div 4 - 4 = 11$
14. $120 - 35 - 48 = 37$
15. $(9 + 8) \times 8 = 136$
16. $12 \times 7 + 21 = 105$
17. $64 \div (34 - 18) = 4$
18. $16 \times 8 \div 4 = 32$
19. $16 + 121 \div 11 = 27$
20. $66 - 32 - 34 = 0$
21. $96 \div 2 \div 4 = 12$
22. $9 \times 2 \times 6 = 108$
23. $(55 + 35) \div 18 = 5$
24. $(100 - 28) \div 8 = 9$

Name: Lesson 9-8 Order of operations

1. $29 + 14 - 17 = 26$
2. $5 \times 6 - 9 = 21$
3. $78 - 9 \times 5 = 33$
4. $99 \div 9 + 13 = 24$
5. $9 \times (11 - 2) = 81$
6. $(14 - 7) \times 13 = 91$
7. $53 + 26 + 55 = 134$
8. $54 \div (12 + 6) = 3$
9. $105 \div 5 - 13 = 8$
10. $22 - 35 \div 7 = 17$
11. $55 + 4 \times 4 = 71$
12. $67 - 32 - 19 = 16$
13. $64 \div (12 - 4) = 8$
14. $8 \times 12 + 35 = 131$
15. $(12 + 24) \times 8 = 288$
16. $128 \div 4 \div 2 = 16$
17. $135 \div 9 \div 5 = 3$
18. $55 - 18 + 22 = 59$
19. $150 \div 5 \times 4 = 120$
20. $9 \times 8 + 71 = 143$
21. $39 + 39 \div 3 = 52$
22. $7 \times 5 \times 9 = 315$
23. $(45 + 27) \div 3 = 24$
24. $(90 - 36) \div 3 = 18$

Name: **Lesson 9-9 Order of operations**

1. $21 + 19 - 6 = 34$
2. $5 \times 5 - 6 = 19$
3. $54 - 9 \times 3 = 27$
4. $51 \div 3 + 18 = 35$
5. $33 + 8 + 7 = 48$
6. $(32 - 18) \times 5 = 70$
7. $12 \times (7 - 6) = 12$
8. $45 \div (8 + 7) = 3$
9. $19 + 7 \times 8 = 75$
10. $81 - 7 \times 6 = 39$
11. $72 \div 6 - 5 = 7$
12. $6 \times (14 + 6) = 120$
13. $(12 + 13) \times 5 = 125$
14. $51 - 12 - 15 = 24$
15. $98 \div (11 - 4) = 14$
16. $6 \times 8 + 44 = 92$
17. $100 \div (33 - 8) = 4$
18. $48 \times 6 \div 2 = 144$
19. $15 + 49 \div 7 = 22$
20. $73 - 28 + 31 = 76$
21. $200 \div 5 \div 8 = 5$
22. $5 \times 5 \times 5 = 125$
23. $(55 + 17) \div 9 = 8$
24. $(54 - 9) \div 5 = 9$

Name: **Lesson 10-1 Rounding numbers**

1. $43 + 12 + 27 = 82$
2. $5 \times 6 - 11 = 19$
3. $44 - 5 \times 7 = 9$
4. $36 \div 6 + 26 = 32$
5. $24 + 27 + 38 = 89$
6. $(22 - 11) \times 11 = 121$
7. $9 \times (13 - 4) = 81$
8. $51 \div (7 - 4) = 17$
9. $35 + 7 \times 8 = 91$
10. $15 - 54 \div 6 = 6$
11. $(17 + 19) \times 7 = 252$
12. $6 \times (7 + 7) = 84$
13. $135 \div 9 + 6 = 21$
14. $92 - 24 - 49 = 21$
15. $111 \div (9 - 6) = 37$
16. $7 \times 7 + 7 = 56$
17. $45 + 36 \div 3 = 57$
18. $9 \times 8 \div 3 = 24$
19. $72 \div 6 \times 8 = 96$
20. $57 - 23 - 19 = 15$
21. $96 \div 3 \div 4 = 8$
22. $8 \times 7 \times 6 = 336$
23. $(22 + 33) \div 5 = 11$
24. $(86 - 8) \div 6 = 13$

Name: **Lesson 10-1 Rounding numbers**

Round the following numbers to the nearest ten.
Example: 547 rounded to the nearest ten is 550.

1. 128 = 130
2. 244 = 240
3. 791 = 790
4. 517 = 520
5. 536 = 540
6. 951 = 950
7. 449 = 450
8. 213 = 210
9. 221 = 220
10. 572 = 570
11. 385 = 390
12. 370 = 370
13. 388 = 390
14. 739 = 740
15. 108 = 110
16. 315 = 320
17. 144 = 140
18. 935 = 940
19. 257 = 260
20. 348 = 350
21. 238 = 240
22. 333 = 330
23. 989 = 990
24. 357 = 360
25. 421 = 420
26. 201 = 200
27. 549 = 550

Name: **Lesson 10-2 Rounding numbers**

Round the following numbers to the nearest ten.
Example: 547 rounded to the nearest ten is 550.

1. 855 = 860
2. 618 = 620
3. 696 = 700
4. 977 = 980
5. 205 = 210
6. 985 = 990
7. 279 = 280
8. 984 = 980
9. 632 = 630
10. 275 = 280
11. 724 = 720
12. 313 = 310
13. 122 = 120
14. 302 = 300
15. 117 = 120
16. 253 = 250
17. 250 = 250
18. 241 = 240
19. 308 = 310
20. 946 = 950
21. 853 = 850
22. 962 = 960
23. 925 = 930
24. 374 = 370
25. 837 = 840
26. 114 = 110
27. 606 = 610

Name: **Lesson 10-3 Rounding numbers**

Round the following numbers to the nearest ten.
Example: 547 rounded to the nearest ten is 550.

1. 696 = 700	2. 638 = 640	3. 125 = 130
4. 701 = 700	5. 929 = 930	6. 450 = 450
7. 882 = 880	8. 674 = 670	9. 467 = 470
10. 377 = 380	11. 799 = 800	12. 286 = 290
13. 118 = 120	14. 552 = 550	15. 817 = 820
16. 451 = 450	17. 338 = 340	18. 981 = 980
19. 979 = 980	20. 219 = 220	21. 492 = 490
22. 698 = 700	23. 622 = 620	24. 404 = 400
25. 544 = 540	26. 881 = 880	27. 229 = 230

Name: **Lesson 10-4 Rounding numbers**

Round the following numbers to the nearest ten.
Example: 547 rounded to the nearest ten is 550.

1. 617 = 620	2. 867 = 870	3. 667 = 670
4. 807 = 810	5. 257 = 260	6. 598 = 600
7. 773 = 770	8. 717 = 720	9. 277 = 280
10. 464 = 460	11. 468 = 470	12. 920 = 920
13. 925 = 930	14. 477 = 480	15. 836 = 840
16. 624 = 620	17. 213 = 210	18. 256 = 260
19. 334 = 330	20. 362 = 360	21. 725 = 730
22. 455 = 460	23. 635 = 640	24. 247 = 250
25. 805 = 810	26. 138 = 140	27. 101 = 100

Name: **Lesson 10-5 Rounding numbers**

Round the following numbers to the nearest ten.
Example: 547 rounded to the nearest ten is 550.

1. 731 = 730	2. 814 = 810	3. 717 = 720
4. 222 = 220	5. 729 = 730	6. 942 = 940
7. 594 = 590	8. 389 = 390	9. 729 = 730
10. 819 = 820	11. 367 = 370	12. 132 = 130
13. 936 = 940	14. 233 = 230	15. 377 = 380
16. 492 = 490	17. 187 = 190	18. 567 = 570
19. 402 = 400	20. 236 = 240	21. 583 = 580
22. 645 = 650	23. 784 = 780	24. 338 = 340
25. 352 = 350	26. 452 = 450	27. 223 = 220

Name: **Lesson 10-6 Rounding numbers**

Round the following numbers to the nearest hundred.
Example: 678 rounded to the nearest hundred is 700.

1. 897 = 900	2. 791 = 800	3. 237 = 200
4. 191 = 200	5. 369 = 400	6. 487 = 500
7. 522 = 500	8. 507 = 500	9. 972 = 1000
10. 327 = 300	11. 487 = 500	12. 318 = 300
13. 962 = 1000	14. 917 = 900	15. 501 = 500
16. 724 = 700	17. 399 = 400	18. 405 = 400
19. 648 = 600	20. 175 = 200	21. 233 = 200
22. 697 = 700	23. 432 = 400	24. 176 = 200
25. 102 = 100	26. 989 = 1000	27. 960 = 1000

Name: **Lesson 10-7** **Rounding numbers**

Round the following numbers to the nearest hundred.
Example: 678 rounded to the nearest hundred is 700.

1. 2227 = 2200	2. 735 = 700	3. 5937 = 5900
4. 1566 = 1600	5. 3512 = 3500	6. 408 = 400
7. 943 = 900	8. 7292 = 7300	9. 155 = 200
10. 887 = 900	11. 4037 = 4000	12. 3852 = 3900
13. 3856 = 3900	14. 405 = 400	15. 2840 = 2800
16. 627 = 600	17. 138 = 100	18. 1107 = 1100
19. 2247 = 2200	20. 1536 = 1500	21. 117 = 100
22. 5564 = 5600	23. 793 = 800	24. 1220 = 1200
25. 610 = 600	26. 4270 = 4300	27. 538 = 500

Name: **Lesson 10-8** **Rounding numbers**

Round the following numbers to the nearest hundred.
Example: 678 rounded to the nearest hundred is 700.

1. 396 = 400	2. 8566 = 8600	3. 108 = 100
4. 892 = 900	5. 711 = 700	6. 779 = 800
7. 4581 = 4600	8. 160 = 200	9. 5158 = 5200
10. 2401 = 2400	11. 7452 = 7500	12. 505 = 500
13. 365 = 400	14. 3572 = 3600	15. 8234 = 8200
16. 529 = 500	17. 226 = 200	18. 9477 = 9500
19. 741 = 700	20. 8030 = 8000	21. 7174 = 7200
22. 8559 = 8600	23. 149 = 100	24. 928 = 900
25. 2516 = 2500	26. 9947 = 9900	27. 607 = 600

Name: **Lesson 10-9** **Rounding numbers**

Round the following numbers to the nearest thousand.
Example: 1785 rounded to the nearest thousand is 2000.

1. 1431 = 1000	2. 4115 = 4000	3. 913 = 1000
4. 5630 = 6000	5. 843 = 1000	6. 8938 = 9000
7. 6557 = 7000	8. 8959 = 9000	9. 7918 = 8000
10. 7749 = 8000	11. 639 = 1000	12. 6927 = 7000
13. 7415 = 7000	14. 3502 = 4000	15. 1713 = 2000
16. 9728 = 10000	17. 2860 = 3000	18. 4358 = 4000
19. 8242 = 8000	20. 922 = 1000	21. 6324 = 6000
22. 8903 = 9000	23. 5187 = 5000	24. 4276 = 4000
25. 4834 = 5000	26. 8124 = 8000	27. 2518 = 3000

Name: **Lesson 10-10** **Rounding numbers**

Round the following numbers to the nearest thousand.
Example: 1785 rounded to the nearest thousand is 2000.

1. 3296 = 3000	2. 35675 = 36000	3. 5673 = 6000
4. 6754 = 7000	5. 7190 = 7000	6. 2145 = 2000
7. 19052 = 19000	8. 86609 = 87000	9. 32494 = 32000
10. 42849 = 43000	11. 7198 = 7000	12. 6630 = 7000
13. 9963 = 10000	14. 2759 = 3000	15. 89430 = 89000
16. 16660 = 17000	17. 41078 = 41000	18. 6544 = 7000
19. 2227 = 2000	20. 77796 = 78000	21. 38186 = 38000
22. 6884 = 7000	23. 2149 = 2000	24. 4517 = 5000
25. 14890 = 15000	26. 3481 = 3000	27. 2592 = 3000

Name: **Lesson 11-1 Finding the average**

Find the sum and average of the following numbers. For example,
The sum of 24 and 48 is $24+48=72$.
The average of 24 and 48 is $\frac{24+48}{2}=\frac{72}{2}=36$.

1. 24, 48 • Sum = 72 • Average = 36
2. 17, 33 • Sum = 50 • Average = 25
3. 42, 26 • Sum = 68 • Average = 34
4. 36, 42 • Sum = 78 • Average = 39
5. 11, 99 • Sum = 110 • Average = 55
6. 33, 53 • Sum = 86 • Average = 43
7. 17, 59 • Sum = 76 • Average = 38
8. 55, 75 • Sum = 130 • Average = 65
9. 86, 68 • Sum = 154 • Average = 77
10. 22, 82 • Sum = 104 • Average = 52
11. 51, 57 • Sum = 108 • Average = 54

Name: **Lesson 11-2 Finding the average**

Find the sum and average of the following numbers. For example,
The sum of 24 and 48 is $24+48=72$.
The average of 24 and 48 is $\frac{24+48}{2}=\frac{72}{2}=36$.

1. 16, 88 • Sum = 104 • Average = 52
2. 86, 58 • Sum = 144 • Average = 72
3. 52, 74 • Sum = 126 • Average = 63
4. 35, 65 • Sum = 100 • Average = 50
5. 53, 11 • Sum = 64 • Average = 32
6. 98, 56 • Sum = 154 • Average = 77
7. 38, 90 • Sum = 128 • Average = 64
8. 27, 27 • Sum = 54 • Average = 27
9. 43, 91 • Sum = 134 • Average = 67
10. 64, 46 • Sum = 110 • Average = 55
11. 15, 93 • Sum = 108 • Average = 54

Name: **Lesson 11-3 Finding the average**

Find the sum and average of the following numbers. For example,
The sum of 24 and 48 is $24+48=72$.
The average of 24 and 48 is $\frac{24+48}{2}=\frac{72}{2}=36$.

1. 83, 51 • Sum = 134 • Average = 67
2. 10, 86 • Sum = 96 • Average = 48
3. 79, 39 • Sum = 118 • Average = 59
4. 84, 36 • Sum = 120 • Average = 60
5. 85, 69 • Sum = 154 • Average = 77
6. 91, 67 • Sum = 158 • Average = 79
7. 16, 90 • Sum = 106 • Average = 53
8. 46, 10 • Sum = 56 • Average = 28
9. 44, 12 • Sum = 56 • Average = 28
10. 42, 68 • Sum = 110 • Average = 55
11. 62, 38 • Sum = 100 • Average = 50

Name: **Lesson 11-4 Finding the average**

Find the sum and average of the following numbers. For example,
The sum of 24 and 48 is $24+48=72$.
The average of 24 and 48 is $\frac{24+48}{2}=\frac{72}{2}=36$.

1. 33, 39 • Sum = 72 • Average = 36
2. 25, 59 • Sum = 84 • Average = 42
3. 55, 27 • Sum = 82 • Average = 41
4. 69, 41 • Sum = 110 • Average = 55
5. 49, 57 • Sum = 106 • Average = 53
6. 27, 47 • Sum = 74 • Average = 37
7. 37, 87 • Sum = 124 • Average = 62
8. 89, 27 • Sum = 116 • Average = 58
9. 46, 68 • Sum = 114 • Average = 57
10. 28, 66 • Sum = 94 • Average = 47
11. 88, 98 • Sum = 186 • Average = 93

Name: **Lesson 11-5 Finding the average**

Find the sum and average of the following numbers. For example,
The sum of 24 and 48 is $24+48=72$.
The average of 24 and 48 is $\frac{24+48}{2}=\frac{72}{2}=36$.

1. 38, 36 • Sum = 74 • Average = 37
2. 86, 12 • Sum = 98 • Average = 49
3. 61, 91 • Sum = 152 • Average = 76
4. 25, 43 • Sum = 68 • Average = 34
5. 44, 76 • Sum = 120 • Average = 60
6. 10, 24 • Sum = 34 • Average = 17
7. 37, 85 • Sum = 122 • Average = 61
8. 47, 81 • Sum = 128 • Average = 64
9. 80, 32 • Sum = 112 • Average = 56
10. 40, 28 • Sum = 68 • Average = 34
11. 30, 66 • Sum = 96 • Average = 48

Name: **Lesson 11-6 Finding the average**

Find the sum and average of the following numbers. For example,
The sum of 24 and 48 is $24+48=72$.
The average of 24 and 48 is $\frac{24+48}{2}=\frac{72}{2}=36$.

1. 89, 91 • Sum = 180 • Average = 90
2. 90, 28 • Sum = 118 • Average = 54
3. 113, 43 • Sum = 156 • Average = 78
4. 69, 57 • Sum = 126 • Average = 63
5. 103, 87 • Sum = 190 • Average = 95
6. 88, 96 • Sum = 184 • Average = 92
7. 52, 56 • Sum = 108 • Average = 54
8. 74, 128 • Sum = 202 • Average = 101
9. 114, 76 • Sum = 190 • Average = 95
10. 94, 118 • Sum = 212 • Average = 106
11. 14, 138 • Sum = 152 • Average = 76

Name: **Lesson 11-7 Finding the average**

Find the sum and average of the following numbers. For example,
The sum of 24 and 48 is $24+48=72$.
The average of 24 and 48 is $\frac{24+48}{2}=\frac{72}{2}=36$.

1. 130, 112 • Sum = 242 • Average = 121
2. 127, 105 • Sum = 232 • Average = 116
3. 117, 139 • Sum = 256 • Average = 128
4. 39, 131 • Sum = 170 • Average = 85
5. 60, 28 • Sum = 88 • Average = 44
6. 19, 59 • Sum = 78 • Average = 39
7. 129, 141 • Sum = 270 • Average = 135
8. 47, 107 • Sum = 158 • Average = 79
9. 54, 146 • Sum = 200 • Average = 100
10. 45, 61 • Sum = 106 • Average = 53
11. 119, 151 • Sum = 270 • Average = 135

Name: **Lesson 11-8 Finding the average**

Find the sum and average of the following numbers. For example,
The sum of 24 and 48 is $24+48=72$.
The average of 24 and 48 is $\frac{24+48}{2}=\frac{72}{2}=36$.

1. 69, 153 • Sum = 222 • Average = 111
2. 86, 104 • Sum = 190 • Average = 95
3. 22, 154 • Sum = 176 • Average = 88
4. 159, 69 • Sum = 228 • Average = 114
5. 61, 55 • Sum = 116 • Average = 58
6. 145, 55 • Sum = 200 • Average = 100
7. 37, 79 • Sum = 116 • Average = 58
8. 63, 151 • Sum = 214 • Average = 107
9. 36, 42 • Sum = 78 • Average = 39
10. 31, 117 • Sum = 148 • Average = 74
11. 99, 81 • Sum = 180 • Average = 90

Name: **Lesson 11-9 Finding the average**

Find the sum and average of the following numbers. For example,
The sum of 24 and 48 is $24 + 48 = 72$.
The average of 24 and 48 is $\frac{24+48}{2} = \frac{72}{2} = 36$.

1. 177, 45 • Sum = 222 • Average = 111
2. 103, 45 • Sum = 148 • Average = 74
3. 90, 118 • Sum = 208 • Average = 104
4. 125, 101 • Sum = 226 • Average = 113
5. 37, 17 • Sum = 54 • Average = 27
6. 94, 144 • Sum = 238 • Average = 119
7. 43, 101 • Sum = 144 • Average = 72
8. 164, 132 • Sum = 296 • Average = 148
9. 102, 130 • Sum = 232 • Average = 116
10. 35, 109 • Sum = 144 • Average = 72
11. 31, 119 • Sum = 150 • Average = 75

Name: **Lesson 11-10 Finding the average**

Find the sum and average of the following numbers. For example,
The sum of 24 and 48 is $24 + 48 = 72$.
The average of 24 and 48 is $\frac{24+48}{2} = \frac{72}{2} = 36$.

1. 158, 68 • Sum = 226 • Average = 113
2. 176, 14 • Sum = 190 • Average = 95
3. 129, 151 • Sum = 280 • Average = 140
4. 92, 146 • Sum = 238 • Average = 119
5. 162, 188 • Sum = 350 • Average = 175
6. 57, 75 • Sum = 132 • Average = 66
7. 171, 123 • Sum = 294 • Average = 147
8. 118, 168 • Sum = 286 • Average = 143
9. 71, 57 • Sum = 128 • Average = 64
10. 34, 144 • Sum = 178 • Average = 89
11. 136, 58 • Sum = 194 • Average = 97

Name: **Lesson 12-1 Counting the total value of coins**

Count the total value of the following coins in cents.
Example: The total value of 1 quarters, 1 dime, 1 nickel, and 1 penny is $25 + 10 + 5 + 1 = 41$ cents.

1. 2 quarters, 1 dime, and 1 penny = 61 cents
2. 1 quarter, 3 dimes, and 2 pennies = 57 cents
3. 2 quarters, 1 nickel, and 3 pennies = 58 cents
4. 1 quarter, 2 dimes, and 1 nickel = 50 cents
5. 2 quarters, 1 dime, and 2 nickels = 70 cents
6. 1 quarter, 3 dimes, and 1 nickel = 60 cents
7. 3 quarters, 1 dimes, and 1 nickel = 90 cents
8. 3 quarters, 1 dime, and 3 pennies = 88 cents
9. 3 quarters, 2 nickels, and 2 pennies = 87 cents
10. 3 quarters, 3 dimes, and 2 pennies = 107 cents
11. 3 quarters, 3 dimes, and 3 nickels = 120 cents

Name: **Lesson 12-2 Counting the total value of coins**

Count the total value of the following coins in cents.
Example: The total value of 1 quarters, 1 dime, 1 nickel, and 1 penny is $25 + 10 + 5 + 1 = 41$ cents.

1. 2 quarters, 2 dimes, and 2 pennies = 72 cents
2. 2 quarters, 4 dimes, and 3 nickels = 105 cents
3. 3 quarters, 1 dime, and 2 nickels = 95 cents
4. 2 quarters, 4 nickels, and 3 pennies = 73 cents
5. 3 quarters, 2 dimes, and 1 nickel = 100 cents
6. 4 quarters, 2 dimes, and 3 nickels = 135 cents
7. 3 quarters, 4 dimes, and 1 penny = 116 cents
8. 3 quarters, 3 nickels, and 3 pennies = 93 cents
9. 4 quarters, 3 dimes, and 4 nickels = 150 cents
10. 4 quarters, 4 dimes, and 3 nickels = 155 cents
11. 2 quarters, 3 nickels, and 1 penny = 66 cents

Name: **Lesson 12-3 Counting the total value of coins**

Count the total value of the following coins in cents.
Example: The total value of 1 quarters, 1 dime, 1 nickel, and 1 penny is $25 + 10 + 5 + 1 = 41$ cents.

1. 3 quarters, 2 nickels, and 2 pennies = 87 cents
2. 2 quarters, 4 dimes, and 3 nickels = 105 cents
3. 4 quarters, 1 nickel, and 3 pennies = 108 cents
4. 4 quarters, 2 nickels, and 1 penny = 111 cents
5. 4 quarters, 4 nickels, and 3 pennies = 123 cents
6. 3 quarters, 4 nickels, and 3 pennies = 98 cents
7. 2 quarters, 3 dimes, and 3 nickels = 95 cents
8. 4 quarters, 3 nickels, and 3 pennies = 118 cents
9. 3 quarters, 3 nickels, and 4 pennies = 94 cents
10. 4 quarters, 3 dimes, and 2 pennies = 132 cents
11. 2 quarters, 3 dimes, and 4 nickels = 100 cents

Name: **Lesson 12-4 Counting the total value of coins**

Count the total value of the following coins in cents.
Example: The total value of 1 quarters, 1 dime, 1 nickel, and 1 penny is $25 + 10 + 5 + 1 = 41$ cents.

1. 4 quarters, 4 dimes, and 4 pennies = 144 cents
2. 4 quarters, 1 dime, and 1 nickel = 115 cents
3. 3 quarters, 3 dimes, and 1 penny = 106 cents
4. 2 quarters, 4 dimes, and 3 pennies = 93 cents
5. 4 quarters, 3 dimes, and 3 pennies = 133 cents
6. 4 quarters, 2 dimes, and 2 nickels = 130 cents
7. 3 quarters, 4 dimes, and 2 pennies = 117 cents
8. 4 quarters, 1 dime, and 4 pennies = 114 cents
9. 3 quarters, 1 dime, and 2 pennies = 87 cents
10. 4 quarters, 3 dimes, and 2 nickels = 140 cents
11. 4 quarters, 2 dimes, and 1 penny = 121 cents

Name: **Lesson 12-5 Counting the total value of coins**

Count the total value of the following coins in cents.
Example: The total value of 1 quarters, 1 dime, 1 nickel, and 1 penny is $25 + 10 + 5 + 1 = 41$ cents.

1. 3 quarters, 4 dimes, and 3 pennies = 118 cents
2. 3 quarters, 3 dimes, and 5 nickels = 130 cents
3. 4 quarters, 3 dimes, and 3 pennies = 133 cents
4. 5 quarters, 3 nickels, and 5 pennies = 145 cents
5. 3 quarters, 5 dimes, and 4 pennies = 129 cents
6. 4 quarters, 5 dimes, and 3 pennies = 153 cents
7. 3 quarters, 5 dimes, and 3 nickels = 140 cents
8. 5 quarters, 4 nickels, and 3 pennies = 148 cents
9. 3 quarters, 2 dimes, and 1 penny = 96 cents
10. 4 quarters, 5 dimes, and 2 pennies = 152 cents
11. 3 quarters, 4 dimes, and 1 nickel = 120 cents

Name: **Lesson 12-6 Counting the total value of coins**

Count the total value of the following coins in cents.
Example: The total value of 1 quarters, 1 dime, 1 nickel, and 1 penny is $25 + 10 + 5 + 1 = 41$ cents.

1. 5 quarters, 1 nickel, and 2 pennies = 132 cents
2. 4 quarters, 5 dimes, and 1 nickel = 155 cents
3. 3 quarters, 4 dimes, and 4 pennies = 119 cents
4. 2 quarters, 3 dimes, and 5 pennies = 85 cents
5. 5 quarters, 4 nickels, and 3 pennies = 148 cents
6. 4 quarters, 4 dimes, and 2 nickels = 150 cents
7. 3 quarters, 5 dimes, and 4 pennies = 129 cents
8. 5 quarters, 5 nickels, and 4 pennies = 154 cents
9. 4 quarters, 3 dimes, and 5 nickels = 155 cents
10. 3 quarters, 1 dime, and 5 pennies = 90 cents
11. 5 quarters, 3 nickels, and 1 penny = 141 cents

Name: Lesson 12-7 Counting the total value of coins

Count the total value of the following coins in cents.

Example: The total value of 1 quarters, 1 dime, 1 nickel, and 1 penny is $25 + 10 + 5 + 1 = 41$ cents.

1. 5 quarters, 5 dimes, and 5 nickels = 200 cents
2. 6 quarters, 1 dime, and 3 pennies = 163 cents
3. 4 quarters, 4 nickels, and 1 penny = 121 cents
4. 5 quarters, 3 dimes, and 3 nickels = 170 cents
5. 6 quarters, 2 dimes, and 5 pennies = 175 cents
6. 5 quarters, 1 nickel, and 2 pennies = 132 cents
7. 5 quarters, 1 dime, and 3 nickels = 150 cents
8. 6 quarters, 3 dimes, and 3 pennies = 183 cents
9. 4 quarters, 2 nickels, and 6 pennies = 116 cents
10. 6 quarters, 6 dimes, and 6 pennies = 216 cents
11. 5 quarters, 4 dimes, and 2 nickels = 175 cents

Name: Lesson 12-8 Counting the total value of coins

Count the total value of the following coins in cents.

Example: The total value of 1 quarters, 1 dime, 1 nickel, and 1 penny is $25 + 10 + 5 + 1 = 41$ cents.

1. 5 quarters, 4 nickels, and 4 pennies = 149 cents
2. 6 quarters, 6 dimes, and 5 pennies = 215 cents
3. 4 quarters, 1 nickel, and 2 pennies = 107 cents
4. 3 quarters, 3 dimes, and 5 nickels = 130 cents
5. 5 quarters, 5 nickels, and 6 pennies = 156 cents
6. 6 quarters, 5 dimes, and 6 pennies = 206 cents
7. 4 quarters, 2 nickels, and 4 pennies = 114 cents
8. 3 quarters, 6 nickels, and 5 pennies = 110 cents
9. 6 quarters, 4 dimes, and 1 penny = 191 cents
10. 2 quarters, 5 nickels, and 2 pennies = 77 cents
11. 6 quarters, 2 dimes, and 3 pennies = 173 cents

Name: Lesson 12-9 Counting the total value of coins

Count the total value of the following coins in cents.

Example: The total value of 1 quarters, 1 dime, 1 nickel, and 1 penny is $25 + 10 + 5 + 1 = 41$ cents.

1. 7 quarters, 1 dime, and 4 pennies = 189 cents
2. 6 quarters, 7 nickels, and 1 penny = 186 cents
3. 5 quarters, 1 dime, and 7 nickels = 170 cents
4. 7 quarters, 3 dimes, and 7 pennies = 212 cents
5. 6 quarters, 6 nickels, and 7 pennies = 187 cents
6. 3 quarters, 2 dimes, and 5 nickels = 120 cents
7. 5 quarters, 3 dimes, and 6 nickels = 185 cents
8. 4 quarters, 4 dimes, and 6 pennies = 146 cents
9. 6 quarters, 5 nickels, and 5 pennies = 180 cents
10. 1 quarter, 7 dimes, and 3 nickels = 110 cents
11. 7 quarters, 6 dimes, and 3 pennies = 238 cents

Name: Lesson 12-10 Counting the total value of coins

Count the total value of the following coins in cents.

Example: The total value of 1 quarters, 1 dime, 1 nickel, and 1 penny is $25 + 10 + 5 + 1 = 41$ cents.

1. 7 quarters, 7 dimes, and 1 penny = 246 cents
2. 6 quarters, 1 dime, and 2 nickels = 170 cents
3. 5 quarters, 7 nickels, and 7 pennies = 167 cents
4. 4 quarters, 6 dimes, and 7 nickels = 195 cents
5. 3 quarters, 6 dimes, and 3 pennies = 138 cents
6. 6 quarters, 2 dimes, and 4 nickels = 190 cents
7. 5 quarters, 6 nickels, and 5 pennies = 160 cents
8. 2 quarters, 3 dimes, and 4 pennies = 84 cents
9. 1 quarter, 7 dimes, and 3 nickels = 110 cents
10. 5 quarters, 5 nickels, and 5 pennies = 155 cents
11. 7 quarters, 2 dimes, and 7 pennies = 202 cents

www.ingramcontent.com/pod-product-compliance
Lightning Source LLC
Chambersburg PA
CBHW080830220526
45467CB00008B/2249
9781717064189